Smart and Functional Soft Materials

Edited by Xufeng Dong

Published in London, United Kingdom

IntechOpen

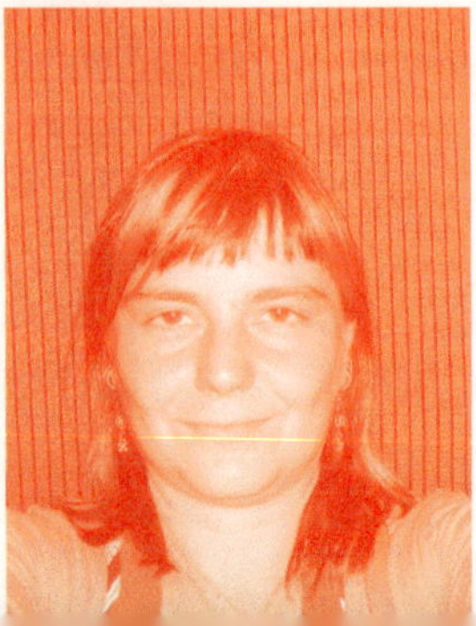

Supporting open minds since 2005

Smart and Functional Soft Materials
http://dx.doi.org/10.5772/intechopen.77491
Edited by Xufeng Dong

Contributors
Safaa Najah Saud, Amir Saud, Mehmet Cabuk, Yangguang Xu, Jennifer Xiaopei Wu, Li Li, Edmundas Kibirkštis, Darius Pauliukaitis, Kęstutis Vaitasius, Narcisa Vrinceanu, Diana Coman, Xufeng Dong

First published in London, United Kingdom, 2019 by IntechOpen
IntechOpen is the global imprint of INTECHOPEN LIMITED, registered in England and Wales, registration number: 11086078, 7th floor, 10 Lower Thames Street, London, EC3R 6AF, United Kingdom
Printed in Croatia

British Library Cataloguing-in-Publication Data
A catalogue record for this book is available from the British Library

Additional hard and PDF copies can be obtained from orders@intechopen.com

Smart and Functional Soft Materials
Edited by Xufeng Dong
p. cm.
Print ISBN 978-1-78984-716-1
Online ISBN 978-1-78984-717-8
eBook (PDF) ISBN 978-1-83962-785-9

For EU product safety concerns:
IN TECH d.o.o., Prolaz Marije Krucifikse Kozulić 3, 51000 Rijeka, Croatia, info@intechopen.com or visit our website at intechopen.com.

Meet the editor

Dr. Xufeng Dong works as a full-time teacher-researcher in materials science. He received his BE, ME, and PhD degrees from the Harbin Institute of Technology, China, in 2003, 2005, and 2009, respectively. He has been with the School of Materials Science and Engineering of Dalian University of Technology as a postdoctoral fellow from 2009 to 2010, as a lecture (2011–2012) and as an associate professor (2013–). Dr. Dong's research interests mainly focus on smart materials, including magnetorheological materials and electrorheological materials. He has published more than 60 technical articles in peer-reviewed journals and conferences.

Contents

Preface

The purpose of this book is to concentrate on recent developments in smart and functional soft materials. The articles collected in this book have been contributed by excellent researchers with long-standing experience in different materials. I hope that the content presented here will be of interest to broader readers, not only scholars in these fields but also people with other disciplines. The book contains seven chapters in two sections: (1) "Smart soft materials" and (2) "Functional soft materials." The book provides detailed and current reviews in these different areas written by experts in their respective fields and will be useful for engineers, scientists, and students who are interested in these areas.

Xufeng Dong, PhD
Associate Professor,
School of Materials Science and Engineering,
Dalian University of Technology,
China

Section 1

Introduction

Chapter 1

Introductory Chapter: The Way to Fulfill Science Fiction

Xufeng Dong

1. Introduction

The history of human civilization is accompanied by the progress of materials. Several hundred thousand years ago, our ancestors went out of Africa, and they began to use stones to make weapons to fight with wild animals or build residence. Several thousand years ago, the early civilized human have mastered the skill to make pottery wares. About 3000 BC, the inhabitant of the Crete Island has entered the Bronze Ware Period. Then there was the Iron Age, and while work efficiency has been greatly improved, battles between different countries became more brutal. Since Bessemer Sir Hery invent the converter steelmaking technology in 1856, the industrial revolution entered the Steel Age, which cause the dramatic progress of civilization. We have used steel to make big machines, cars, airplanes, huge ships, skyscrapers, long-span bridges, rocket ships, and so on. During the past 100 years, we have synthesized a large number of polymer materials, which provided us with various tools, wares, clothes, etc. with good properties but low weight and low cost. Recently, smart soft materials that could response to an external stimulus (such as electric field, magnetic field, sound, light, temperature, pH, and so on) as well as functional soft materials that are electronic conductive, magnetic conductive or thermal conductive have attracted considerable attention. They have application potentials in various fields, including artificial muscles, soft wearable materials, sensors, actuators, soft batteries, soft screens, etc. To some extent, they are the way to fulfill most of the "black technology" that described in some science fictions. Following two sections will give a brief introduction on several main smart soft materials and functional soft materials.

2. Smart soft materials

2.1 Shape memory polymers

Shape-memory polymers (SMPs) are those polymers that have the ability to "memorize" a macroscopic (permanent) shape, be manipulated and "fixed" to a temporary and dormant shape under specific conditions of temperature and stress, and then later relax to the original, stress-free condition under thermal, electrical, or environmental command. This relaxation is associated with elastic deformation stored during prior manipulation [1]. Shape-memory polymers have aroused great attention from scientists and engineers due to their capacity to remember two shapes at different conditions. Potential applications for SMPs exist in almost every area of daily life, such as self-repairing auto bodies, kitchen utensils, switches, sensor, intelligent packing and tools, and so on [2]. Other potential applications are drug delivery [3], biosensors, biomedical devices, microsystem components, and

IntechOpen

smart textile [4]. Since polymer can be made biodegradable, they can be used as short-term implants where removal by surgery can be avoided.

2.2 Dielectric elastomers

Dielectric elastomers (DEs) are an electronically actuated polymer material. They are the typical investigated electro-active polymers (EAPs) and consist of a low stiffness elastomer placed between two conductive electrodes [5]. The shape or volume of DEs can be changed by external electrical stimulation and can be restored to their original shape or volume when the electrical stimulation is withdrawn. DEs have the characteristics of large actuation strain, high bandwidth, high energy density and high conversion efficiency, good environmental adaptability, long fatigue life, and excellent bionic performance. A feature of DEs is that they can not only convert electrical into mechanical energy actuators (DEAs) but also transduce mechanical into electrical energy generators (DEGs) [6]. In the future, DEs are to be used reliably in applications that include soft robotics, medical devices, artificial muscles, and electronic skins [7].

2.3 Electrorheological composites

Electrorheological (ER) composites have been considered as one kind of intelligent materials, and their rheological properties can be instantaneously and reversibly changed by the applied electric field [8]. As one pioneering example of ER composites, electrorheological fluids (ERFs) are commonly composed of polarizable particles and nonconducting liquid. Their rheological properties including viscosity, shear stress, and yield stress can be switched reversible when an electric field is applied [9]. Due to the special properties, ER fluids can be applied in various fields, such as clutches, brakes, accurate polishing, tactile displays, and sensors. As another kind of ER composites, electrorheological elastomers (EREs) also usually filled with dielectric particle as the dispersion phase, which is quite similar to ERFs. However, rubbers or gels are adopted as the matrix in EREs, which enables them to overcome the defects of particle aggregation and sedimentation of ERF [10]. Under the external electric field, EREs can present a reversible change in storage modulus, loss modulus, and loss factors. Owing to these unique advantages of EREs, they are thought of as promising materials for building base isolation, vibration reduction, noise control in mechanical equipment and electro-active actuators [11].

2.4 Magnetorheological composites

Magnetorheological (MR) composites are a kind of smart soft materials whose rheological properties can be actively changed by applying magnetic fields [12]. Magnetorheological fluids (MRFs) and magnetorheological elastomers (MREs) are two main kinds of magnetorheological composites. Magnetorheological fluids are composed by dispersing magnetizable particles in non-magnetizable fluids. Their rheological parameters such as viscosity and yield stress strongly depend on magnetic field [13]. They can be used to fabricate smart dampers, clutches, actuators, human muscle simulators in various fields. The main difference between the magnetorheological elastomers and the magnetorheological fluids is the fluids-matrix that is replaced by polymer-matrix, in most case is elastomers. Compared with MR fluids, MR elastomers present magnetic field dependent dynamic viscoelasticity, including storage modulus, loss modulus, and loss factors [14]. Such unique properties make them to be employed in a wide range of engineering applications such as suspension bushings, seismic protection, engine mounts, adaptively tuned vibration absorbers, and stiffness tunable isolators [15].

2.5 Field responsive gels

Field responsive gels (smart hydrogels) are smart gels which are capable of responding to external stimuli by changing their physic-chemical properties, such as volume, water content, permeability, and hydrophobicity [16]. Field responsive gels take advantages of the hydrophilic and the high swelling ratio of conventional hydrogels and the environmental responsive properties of functional groups or nanoparticles. With an array of triggering mechanisms, including pH, temperature, external chemicals, light, electrical fields, and shear stresses, these field responsive gels enable precise control over fundamental material properties, such as swelling, porosity, viscosity, physical structure, and modulus [17]. With this level of external control, numerous applications within medical and industrial fields have moved into the realm of possibility, such as well-controlled drug delivery, inexpensive and accurate biosensors, artificial muscles, smart films/matrices for tissue engineering, immobilization of enzyme and protein, and adsorption of heavy metals [18].

3. Functional soft materials

3.1 Magnetic responsive polymer composites

Magnetic responsive polymer composites are a kind of functional soft materials that respond to weak magnetic stimuli (static or alternating magnetic field) with a significant effect (e.g., movement, heat generation, magnetic or optical signal) [19]. To date, using composites of polymer matrix and magnetic fillers (micro/nanoparticles) is the most elegant and efficient way to obtain magnetic responsive polymer materials exhibiting high amplitude magneto-response. Magnetic responsive polymer composites can be divided into three categories according to their resulting properties: (i) magnetically driven deformation soft materials, including ferrogels, magneto-active or magnetorheological elastomers, which could be deformed (stretching, bending and rotation) in a controlled manner in homogeneous fields or gradients [20]; (ii) magnetically guided materials for magnetic separation and magnetic targeting in the field of biomedical applications, ranging from drug and cell delivery to diagnostic purposes [21]; and (iii) magnetically actuated thermo-responsive materials for remotely controlled drug delivery and shape shifting [22]. Due to their attractive controllable properties, magnetic responsive polymer composites have potential applications in the biomedical, coatings, microfluidics, and microelectronics fields.

3.2 Electronic conductive polymers

Electronic conductive polymers (ECPs) which are polymers with highly π-conjugated polymeric chains take the attractive advantages with the properties of both conventional polymers and electronic properties of metals or semiconductors. Typical conductive polymers consist of polyacetylene, polyaniline, polypyrrole, and polythiophene. Recently, these ECPs have been paid much more attention in polymer research than conventional polymers because of its unique properties, such as excellent conductivity, large surface areas, and shortened pathways for charge/mass transports, similar to that of the large aspect ratio and multifunctional conductive fillers. Different synthetic strategies and issues regarding morphological control methods, such as in-situ polymerization, hard template methods, heteroatom-doping, and so on, have been explored and developed to prepare various conductive polymer nanostructures, which is beneficial to the development of high-performance devices based on these typically nanostructured conductive polymers [23]. With rationally and desirably designed nanostructure, as a result, these ECPs exhibit excellent electrical, mechanical, and optical properties. Especially, the conductivity of ECPs can be tuned in a wide range up to 10^{-4} S cm^{-1}, although that of neutral conjugated polymers is rather low, usually in the range between 10^{-10} and 10^{-5} S cm^{-1}. Compared with other conventional polymers, the electrical properties of ECPs result from their conductive networks. The electrical properties can be enhanced by the addition of conductive filler, although this procedure leads to poor process-ability and weak mechanical properties [24]. Owing to simple processing tunable properties and easy control of its structure and morphology, ECPs used as one of the most promising candidates in research activities have been studied in wide range of applications, such as energy conversion and storage, sensors, actuators, and biomedical devices [25].

3.3 Thermal conductive polymers

With the development of all classifications of polymers, polymers with high thermal conductivity are of great interest in thermal management systems. Thermal conductive polymers are composed by polymers and filled composites [26]. Polymers have the advantages of good process ability, light weight, low water absorption, high electrical resistivity, high voltage breakdown strength, corrosion resistance, and low cost [27]. Availability of these polymers can expand the plastics industry by partially replacing metals and ceramics in heat transfer devices and systems leading to energy and cost savings. Thermal conductive polymers possess agreeable thermal conductive property. The thermal conductivity of a polymer depends greatly on its morphology and the structure of chains including backbone bonds and side chains, and the inter-chain coupling [28]. What's more, some factors also influence the thermal conductivity of polymers, such as crystallinity, crystal form, orientation of polymer chains, and orientation of ordered domains in both thermoplastics and thermosets were addressed. As for thermal conductive polymers, structure and properties of polymer and fillers (carbon nanotubes, metal particles, and ceramic particles), morphology of the composites, the interaction of polymers and fillers determine the thermal conductivity.

3.4 Bio-medical polymers

Bio-medical polymers as a material intended to interface with biological systems to evaluate, treat, augment or replace any tissue, organ or function of the body, and boundaries for the use of bio-medical polymers are still expanding [29]. The design of new bio-medical polymers is now focused on mimicking many functions of the extracellular matrices of body tissues, as these can regulate host responses in a well-defined manner [30]. However, developing bio-medical polymers with characteristics that could appropriately regenerate tissue or replace the native tissue is a major challenge in this field to date. Recently, naturally derived polymers, such as collagen, alginate, chitosan, and cellulose, have been regaining much attention owing to their inherent biodegradation and biocompatibility, and both physical (roughness, mechanical strength, hydrophobicity, porosity, etc.) and chemical (functionality, drugs, biomolecules, genes, etc.) properties have been modified and their effectiveness were evaluated in performing the desired application, which includes wound healing, drug delivery, tissue engineering applications, and so on [31].

3.5 Semipermeable membrane

A semipermeable membrane is a porous membrane that allows selective diffusion of typical molecules and ions, rather than bio-macromolecules, through the substance [32]. The semipermeable membrane can be a biofilm or a physical membrane, such as animal bladder membrane, casing, eggshell membrane or an artificial semipermeable membranes as cellophane, collodion membranes, and so on [33]. Whether the substance can pass through the semipermeable membrane primarily depends on the differences in concentration on both sides of the membrane. Generally, it only diffuses from one side of high concentration to the side of low concentration. Besides, it also depends on the particle size of the substance that only particles in smaller diameters than the pore size of semipermeable membrane can pass freely. The semipermeable membrane is mainly used for reverse osmosis and ultrafiltration in membrane separation technology [34]. During the past few decades, semipermeable membranes have been investigated and applied in several different applications, including desalination of saline water, wastewater treatment plants, gas separation plants, fuel cells, etc. The research and applications of semipermeable membranes that transfer heat and water vapor in energy exchangers for building, heating, ventilation, and air-conditioning systems are a new development [35].

Author details

Xufeng Dong
School of Materials Science and Engineering, Dalian University of Technology, Dalian, China

*Address all correspondence to: dongxf@dlut.edu.cn

References

[1] Liu C, Chun SB, Mather PT, et al. Chemically cross-linked polycyclooctene: Synthesis, characterization, and shape memory behavior. Macromolecules. 2002;**35**(27):9868-9874

[2] Jeong HM, Song JH, Lee SY, et al. Miscibility and shape memory property of poly(vinyl chloride)/thermoplastic polyurethane blends. Journal of Materials Science. 2001;**36**(22):5457-5463

[3] Metcalfe A, Desfaits A, Salazkin I, et al. Cold hibernated elastic memory foams for endovascular interventions. Biomaterials. 2003;**24**(3):491-497

[4] Ratna D, Karger-Kocsis J. Recent advances in shape memory polymers and composites: A review. Journal of Materials Science. 2008;**43**(1):254-269

[5] Tang C, Li B, Fang HB, et al. A speedy, amphibian, robotic cube: Resonance actuation by a dielectric elastomer. Sensors and Actuators, A: Physical. 2018;**270**:1-7

[6] Pelrine R, Kornbluh R, Pei QB, et al. High-speed electrically actuated elastomers with strain greater than 100%. Science. 2000;**287**(5454):836-839

[7] Zhang Y, Ellingford C, Zhang R. Electrical and mechanical self-healing in high-performance dielectric elastomer actuator materials. Advanced Functional Materials. 2019;**29**:1808431

[8] Gao L, Zhao X. Electrorheological behaviors of barium titanate/gelatin composite hydrogel elastomers. Journal of Applied Polymer Science. 2004;**94**(6):2517-2521

[9] Niu C, Dong X, Zhao H, et al. Properties of aniline-modified strontium titanyl oxalate-based electrorheological suspension. Smart Materials and Structures. 2014;**23**(7):075018

[10] Shen R, Wang X, Lu Y, et al. Polar-molecule-dominated electrorheological fluids featuring high yield stresses. Advanced Materials. 2009;**21**(45):4631-4635

[11] Niu C, Dong X, Qi M. Enhanced electrorheological properties of elastomers containing TiO2/urea core–shell particles. ACS Applied Materials & Interfaces. 2015;**7**(44):24855-24863

[12] Vicente J, Klingenberg DJ, Hidalgo-Alvarez R. Magnetorheological fluids: A review. Soft Matter. 2011;**7**:3701-3708

[13] Tong Y, Dong X, Qi M. High performance magnetorheological fluids with flower-like cobalt particles. Smart Materials and Structures. 2017;**26**:025023. DOI: 10.1088/1361-665X/aa57cc

[14] Tong Y, Dong X, Qi M. Improved tunable range of the field-induced storage modulus by using flower-like particles as the active phase of magnetorheological elastomers. Soft Matter. 2018;**14**:3504-3509. DOI: 10.1039/C8SM00359A

[15] Wang Q, Dong X, Li L, Ou J. Study on an improved variable stiffness tuned mass damper based on conical magnetorheological elastomer isolators. Smart Materials and Structures. 2017;**26**:105028

[16] Wu HQ, Wang C. Biodegradable smart nanogels: A new platform for targeting drug delivery and biomedical diagnostics. Langmuir. 2016;**32**:6211-6225. DOI: 10.1021/acs.langmuir.6b00842

[17] Merino S, Martín C, Kostarelos K, Prato M, Vázquez E. Nanocomposite hydrogels: 3D polymer–nanoparticle

synergies for on-demand drug delivery. ACS Nano. 2015;**9**(5):4686-4697. DOI: 10.1021/acsnano.5b01433

[18] Koetting MC, Peters JT, Steichen SD, et al. Stimulus-responsive hydrogels: Theory, modern advances, and applications. Materials Science & Engineering R: Reports. 2015;**93**:1-49. DOI: 10.1016/j.mser.2015.04.001

[19] Thévenot J, Oliveira H, Sandre O, et al. Magnetic responsive polymer composite materials. Chemical Society Reviews. 2013;**42**(17):7099-7116

[20] Nguyen VQ, Ahmed AS, Ramanujan RV. Morphing soft magnetic composites. Advanced Materials. 2012;**24**(30):4041-4054

[21] Reddy LH, Arias JL, Nicolas J, et al. Magnetic nanoparticles: Design and characterization, toxicity and biocompatibility, pharmaceutical and biomedical applications. Chemical Reviews. 2012;**112**(11):5818-5878

[22] Brazel CS. Magnetothermally-responsive nanomaterials: Combining magnetic nanostructures and thermally-sensitive polymers for triggered drug release. Pharmaceutical Research. 2009;**26**(3):644-656

[23] Shi Y, Peng L, Ding Y, Zhao Y, Yu G. Nanostructured conductive polymers for advanced energy storage. Chemical Society Reviews. 2015;**44**:6684-6696. DOI: 10.1039/C5CS00362H

[24] Deng H, Lin L, Ji M, Zhang S, Yang M, Fu Q. Progress on the morphological control of conductive network in conductive polymer composites and the use as electroactive multifunctional materials. Progress in Polymer Science. 2014;**39**:627-655. DOI: 10.1039/C3EE40997J

[25] Zhao Y, Liu B, Pan L, Yu G. 3D nanostructured conductive polymer hydrogels for high-performance electrochemical devices. Energy & Environmental Science. 2013;**6**: 2856-2870. DOI: 10.1039/C3EE40997J

[26] Afrin R, Jahir H, Abbas A, et al. Review of polymers for heat exchanger applications: Factors concerning thermal conductivity. Applied Thermal Engineering. 2017;**113**:1118-1127

[27] Chen H, Ginzburg VV, Yang J, et al. Thermal conductivity of polymer-based composites: Fundamentals and applications. Progress in Polymer Science. 2016;**59**:41-85

[28] Huang C, Qian X, Yang R. Thermal conductivity of polymers and polymer nanocomposites. Materials Science and Engineering R. 2018;**132**:1-22

[29] Dasgupta Q, Madras G, Chatterjee K. Biodegradable polyol-based polymers for biomedical applications. International Materials Review. 2018;**64**:288-309. DOI: 10.1080/09506608.2018.1505066

[30] Chaudhuri O, Gu L, Klumpers D, Darnell M, Bencherif SA, Weaver JC, et al. Hydrogels with tunable stress relaxation regulate stem cell fate and activity. Nature Materials. 2016;**15**: 326-336. DOI: 10.1038/NMAT4489

[31] Molaei A, Yousefpour M. Preparation of Chitosan-based nanocomposites and biomedical investigations in bone tissue engineering. International Journal of Polymeric Materials and Polymeric Biomaterials. 2019;**68**(12):701-713. DOI: 10.1080/00914037.2018.1493683

[32] Sinha S, Jing H, Das S. Charge inversion and external salt effect in semi-permeable membrane electrostatics. Journal of Membrane Science. 2017;**533**:364-377. DOI: 10.1016/j.memsci.2017.03.049

[33] Tao F, Zhang Y, Wang B, Zhang F, Chang X, Fan R, et al. Graphite powder/

semipermeable collodion membrane composite for water evaporation. Solar Energy Materials & Solar Cells. 2018;**180**:34-45. DOI: 10.1016/j.solmat.2018.02.014

[34] Li D, Yan Y, Wang H. Recent advances in polymer and polymer composite membranes for reverse and forward osmosis processes. Progress in Polymer Science. 2016;**61**:104-155. DOI: 10.1016/j.progpolymsci.2016.03.003

[35] Ge G, Mahmood G, Moghaddam D, Moghaddam D, Simonson C, Besant R, et al. Material properties and measurements for semi-permeable membranes used in energy exchangers. Journal of Membrane Science. 2014;**453**:328-336. DOI: 10.1016/j.memsci.2013.11.013

Section 2

Smart Soft Materials

Chapter 2

Research on Shape Memory Alloys and Magnetorheological Fluids for Use in Pneumatic Actuators

Edmundas Kibirkštis, Darius Pauliukaitis and Kęstutis Vaitasius

Abstract

Adaptive vibrating actuators operating under autovibration regime were developed using materials with shape memory (SM). Mechanical characteristics of such type vibroactuators can be adjusted by changing geometrical parameters of chamber or throttle. It allows to expand the application of pneumatic vibroexciters and use them in different technological processes. Pneumatic vibroactuator elements that consist of magnetorheological (MR) fluids are proposed for micro-throttle and microcompressor design. The main properties of used MR materials and theoretical background of their mechanical and physical parameters are presented, some results of experimental and theoretical research of actuators are introduced and conclusions of the performed research are formulated.

Keywords: pneumatic actuator, magnetorheological fluid, microcompressor, micro choke, shape memory element

1. Introduction

Smart materials can be used in various technologies, including the structures of pneumatic actuators on air cushions. These actuators may be applied in different technological devices: in vehicle suspensions, rotary mechanisms, vibrating damping systems, various medical devices, etc. Some properties of actuators with elements of shape memory alloy and magnetorheological fluid are analysed in this study.

2. Theoretical background of the use of materials containing shape memory elements in pneumatic actuators

Functional capabilities of pneumatic actuators [1] can be extended via the variations of their ignition chamber volume or the geometrical parameters of their choke [2].

The structure of any pneumatic actuator containing a shape memory (SM) element, i.e. a variable volume chamber, is provided in **Figure 1**. Such transducers may be used in vibratory machinery, pneumatic shock-absorber structures and regulatory, guidance, positioning and other equipment.

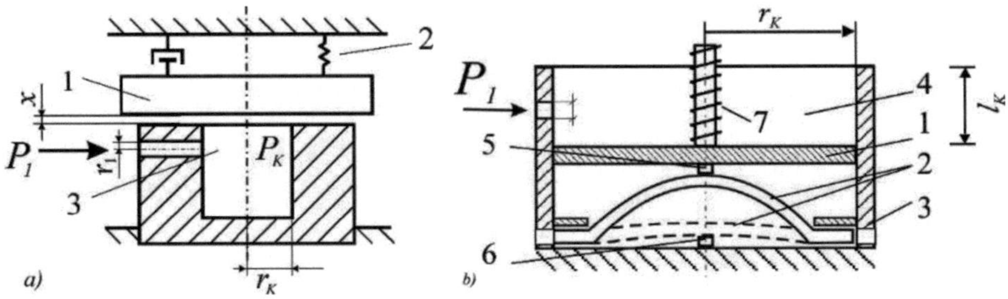

Figure 1.
Schemes of pneumatic vibroexciters constructions, (a) simplified scheme of vibrodrive: 1—oscillatory mass; 2—spring; 3—exciting chamber of vibrodrive; P_1*—supplied gas pressure;* P_K*—pressure inside chamber;* x*—move of oscillatory mass;* r_1*—radius of the air supply channel; and* r_K*—chamber radius. (b) Simplified scheme with element made from smart material: 1—piston; 2—arc-shaped plate with SM material; 3—frame, 4—exciting chamber; 5,6—cooling and heating elements, respectively; 7—spring that pushes piston 1 to the plate 2;* r_K*—exciting chamber radius; and* l_K*—chamber height.*

Any ignition chamber of a pneumatic actuator consists of transmission (**Figure 1b**), which is made of plunger *1*, SM arch element *2*, cooler *5* and heating elements *6*. SM transmission element *2* is situated in case *3*. Transmission *2* contacts heating *6* and cooling *5* elements. Its plate *2* is being heated, and the heat is being transmitted to the arch element. When the arch material reaches the reverse martensitic transformation temperature, it begins to incline in the opposite direction, and when it reaches the opposite position, the system starts cooling. When the arch material reaches temperature which is lower than its phase transformation temperature, the arch escapes to the backwords direction.

By experimental study of shape memory alloys, it was found that temperature range is about 10–16°C when direct and reverse martensitic transformation happens with SM material (NiTi alloy). So, it is important to mention, in a separate case, the system shown in **Figure 1b** without complex feedback control elements 5 and 6.

Based on previous research works [3–5], the plate containing the SM material (nickel and titanium alloy) has been researched as a uniform plate without its decomposition into the core and the outer layer although they could be described as having different deformation properties. Those works offered simplified methods for the calculation of deformations in plates containing SM material. Suppose the plate is situated on a horizontal axis and one or both of its ends are fixed. Its centre is being affected by force which is directed along vertical axis. If the effect of transverse forces that appear during the deformation is ignored, a homogenous bending of the plate in the direction of *Y* axis can be described using the equation of Euler:

$$EJ\frac{d^2y}{dx^2} + Py\left[1 + \left(\frac{dy}{dx}\right)^2\right]^{\frac{3}{2}} = 0 \tag{1}$$

where *E* refers to the Young's modulus, *J* refers to inertia moment and *P* refers to force which affects the centre of the plate. If both ends of the sample are fixed, marginal conditions of Eq. (1) will be as follows:

$$y(0) = y(l) = 0 \tag{2}$$

where *l* refers to the plate length.

Properties of SM material used are presented in **Table 1**.

The number of experiments led to drawing the curve which shows the correlation between mechanical stress with temperatures in nickel and titanium alloy plate

(**Figure 2**). Based on theoretical assumptions contained in the mentioned works [3, 7–17], such calculations can be also used in chains containing SM elements. While heating the plate up to the austenitic temperature, the sinusoidal bending will be obtained:

$$a \sin \frac{\pi \xi}{l} \tag{3}$$

where a refers to the maximum bending of the plate.

Taking into account Eq. (1), the close solution of bending in the direction of Y axis in the sample whose both ends are fixed can be found [3, 7]:

$$y(\xi) = \frac{2^{\frac{3}{2}} l}{\pi} \sqrt{\frac{\sigma - \sigma_1}{\sigma_1}} \sin \frac{\pi \xi}{l} + o\left(\sqrt{\sigma - \sigma_1}\right) \tag{4}$$

where the minimum marginal stress is described as

$$\sigma_1 = \frac{\pi^2 E J}{l^2 S} \tag{5}$$

Martensite start temperature, M_s	18.4°C
Martensite finish temperature, M_f	9°C
Austenite start temperature, A_s	34.5°C
Austenite finish temperature, A_f	49°C
Transformation constant, C_M	0.008 GPa/°C
Transformation constant, C_A	0.0138 GPa/°C
Transformation constant, σ_s^{cr}	0.1 GPa
Transformation constant, σ_f^{cr}	0.17 GPa

Table 1.
Properties of NiTi material [6].

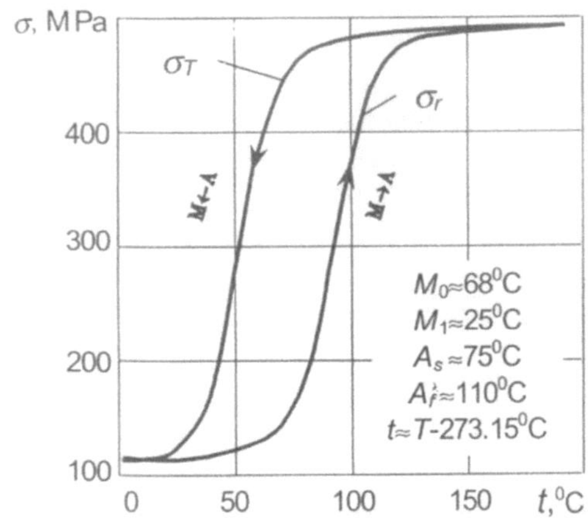

Figure 2.
Correlation between mechanical stress and temperature in nickel and titanium alloy plate: σ_r—reaction stress; σ_T—martensitic transformation stress; M_o and M_1—direct martensitic transformation (A- > M) temperatures, initial and terminal, respectively; and A_o and A_1—reverse martensitic transformation (M- > A) temperatures, initial and terminal, respectively [3].

and the arched axis is described by a sinusoid

$$y(\xi) = \frac{2^{\frac{3}{2}} l}{\pi} \sqrt{\frac{\sigma - \sigma_1}{\sigma_1}} \sin \frac{\pi \xi}{l} \tag{6}$$

while inertia moment of the plate section in regard to neutral axis is equal to

$$J = \frac{bh^3}{12} \tag{7}$$

If $S = bh$ is the plate section area (**Figure 3**), E is an elasticity module of nickel and titanium alloy (app $E = 15.65 \times 10^{10}$ Pa).

The force acting upon the plate with SM effect can be defined by the formula

$$F = \begin{cases} SE_{ad}\alpha(T - T_0), & when\ T_0 \leq T < A_0 \\ S\sigma_r, & when\ A_0 \leq T < A_1 \\ SE_{ad}\alpha(T - A_1), & when\ T > A_1 \end{cases} \tag{8}$$

where T, α and E_{ad} are temperature, thermal expansion coefficient and adiabatic Young's modulus, respectively. By Eqs. (5) and (8), the smallest critical force, when temperature changes between initial and final reverse martensitic transformation temperatures ($A_0 < T < A_1$), will have the following expression:

$$F_1 = \frac{\pi^2 EJ}{l^2} \tag{9}$$

When the temperature varies between A_0 and A_1, the plate regains its present sinusoidal shape. When the temperature approaches critical temperature T_K, corresponding critical stress σ_K, the plate acquires the linear shape.

Pneumatical vibroexciter with exciting chamber radius $r_K = 15 \times 10^{-3}$ m and height $l_K = 20 \times 10^{-3}$ m has been analysed in the numerical research. A smart material element in exciting chamber has used a titanium-nickel plate whose length $l = 25 \times 10^{-3}$ m, height $h = 1.5 \times 10^{-3}$ m and width $b = 20 \times 10^{-3}$ m. By the experimental study [3], approximate Young's modulus for NiTi alloy is $E = 15.65 \times 10^{10}$ Pa, initial and final, respectively, reverse martensitic transformation temperatures $A_0 = 75°C$ or $A_1 = 110°C$.

According to Eqs. (3)–(9), a programme simulating the bending of plate has been proposed. The plate axis bending shapes in the direction of Y axis according to the plate length under the various reaction stresses σ_r are shown in **Figure 4a**.

The numerical researches have showed that a middle point of plate moves within 0.19×10^{-3} m in the direction of Y axis when the reaction stress is equal $\sigma_r = 463.4$ MPa. When the reaction stress is $\sigma_r = 464.0$ MPa, the maximum displacement of the plate middle point is about 0.91×10^{-3} m. According to the alloy

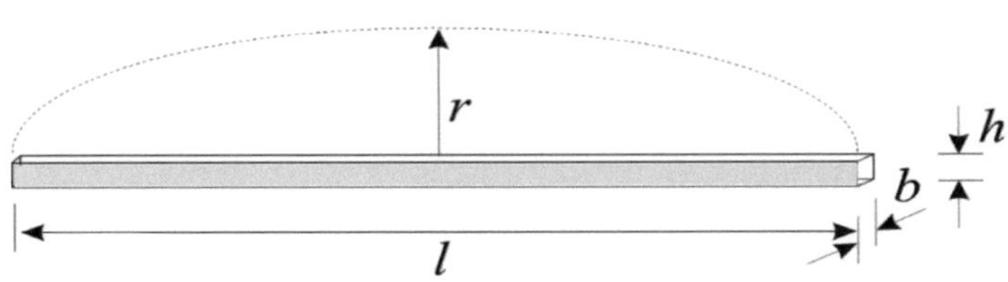

Figure 3.
Scheme of a nickel and titanium alloy plate.

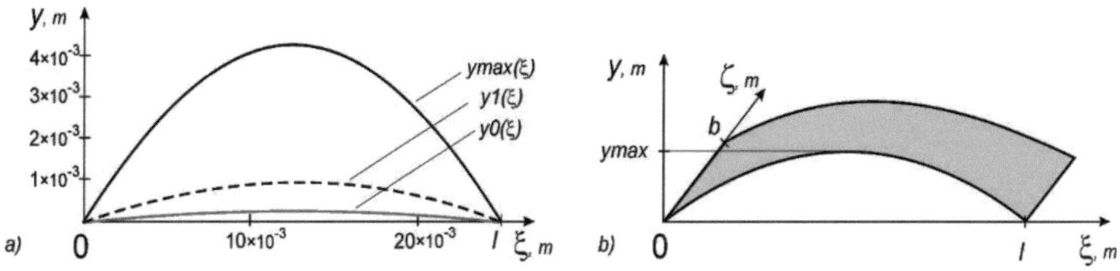

Figure 4.
Deformation of the NiTi plate with the fastened ends: (a) shapes of bended plate under various stresses, curve y0 (ξ), when σ_r = 463.4 MPa, the maximum deflection of bended plate is 0.19 × 10^{-3} m; curve y1(ξ) when σ_r = 464.0 MPa, the maximum deflection of bended plate is 0.91 × 10^{-3} m; and when curve ymax(ξ) when σ_r = 480.0 MPa, the maximum deflection of bended plate is 4.26 × 10^{-3} m; (b) view of bended plate.

characteristics presented in the paper [3], the maximum stress could be σ_r = 480.0 MPa, and maximum deflection could reach the value 4.26 × 10^{-3} m. The critical stress was estimated to be σ_K = 321 MPa.

When stress exceeds critical stress (temperature varies between A_0 and A_1), the sample acquires a sinusoidal shape. It happens when stress exceeds critical margin:

$$\sigma > \left(1 + \frac{\pi^2 a^2}{2^3 l^2}\right)\sigma_K \tag{10}$$

Inequality (10) comes from Eqs. (3) and (4).

When temperature gets close to its critical margin t_K °C and the stress gets close to its critical level respectfully σ_K, bending of the sample decreases and obtains the shape of a line.

The area of the arched plate will be

$$S_B = \frac{L}{2} b \tag{11}$$

where L refers to the parameter of ellipse which is found having entered the complete elliptic interval of level 2 into Legendre form and e refers to elliptic eccentricity:

$$L = 4\frac{l}{2}\int_0^{\frac{\pi}{2}} \sqrt{1 - e^2 \sin\psi d\psi} \tag{12}$$

According to literature [4], linear deformations of the plate can be calculated

$$\varepsilon = \frac{S_B/2 - lb}{lb} \tag{13}$$

Having evaluated methods of identifying bending of the plate under various stresses (Eqs. (4)–(6)) and the deformations under the identified bending (Eqs. (11)–(13)), the authors have developed the software for calculating maximum bending values in plates containing SM materials when their critical stress is known, with additional capability of predicting the shape and the maximum force of bending in the plate. The obtained results are presented in **Figure 5**.

SM materials possess two rather distinctive properties: shape memory effect and superelasticity. Further, we will analyse a mathematical model aimed at the calculation of energy absorbed by the plate containing SM elements while it is being bent.

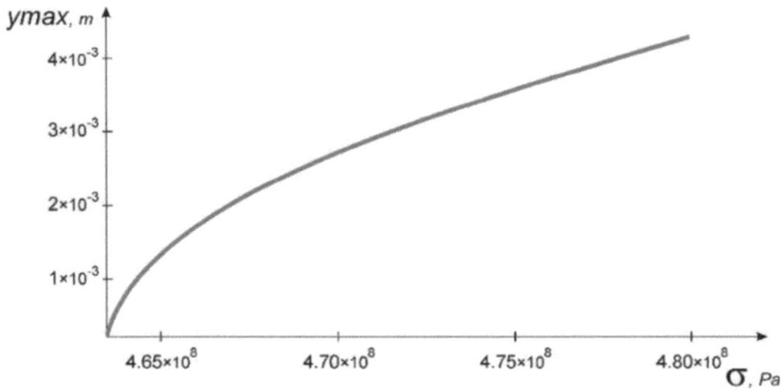

Figure 5.
Dependence of deflection of NiTi plate with the fastened ends on the various reaction stresses.

Further on, we will identify the correlation between maximum bending of the sample and the force which affects it in heating and cooling phases. The overall deformation of the sample ε is equal to the sum of elastic deformation ε^e and phase transition deformation ε^{tr}:

$$\varepsilon = \varepsilon^e + \varepsilon^{tr} \tag{14}$$

According to the Bernoulli equation, the deformation can be expressed as follows [5]:

$$\varepsilon = \frac{2y}{L^2}(\delta_0 - \delta) \tag{15}$$

where δ_0 refers to initial bending and δ refers to bending after the effect of force F.

Elastic deformation ε^e can be expressed as

$$\varepsilon^e = \frac{\sigma}{E} \tag{16}$$

where σ refers to stress which occurs after the effect of force F, while E refers to Young's modulus.

The experiments [5] allowed identifying that deformation which occurs in the transition of phases which is directly proportional to martensitic fraction z_σ:

$$\varepsilon^{tr} = \gamma z_\sigma \tag{17}$$

where γ refers to maximum superelastic deformation.

The sum bending of the plate can be stated as follows:

$$\delta = \delta_0 + \delta_{SMA} + \delta_{EXT} \tag{18}$$

where δ_0 refers to initial bending, δ_{SMA} refers to bending which occurs in the transition of phases and δ_{EXT} refers to bending after the effect of force F.

Final expressions of bending for the process of heating and cooling are stated as follows:

$$\delta = \delta_0 + \frac{3a_2\gamma L^2}{2h}\left(T - A_f^1\right) + \frac{4L^3}{bEh^3}F \text{ (heating)} \tag{19}$$

$$\delta = \delta_0 + \frac{3a_2\gamma L^2}{2h}\left(T - M_f^0 - \frac{1}{a_2}\right) + \frac{4L^3}{bEh^3}F \text{ (cooling)} \tag{20}$$

where a_2 is a constant (measure unit K^{-1}), A^1_f is temperature of terminating the austenitic phase and M^0_f is temperature of terminating the martensitic phase.

Equations (19) and (20) make clear the existence of direct correlation between the temperature and the bending of the plate which describes its thermomechanical behaviour.

Eqs. (19) and (20) can be used for the calculation of bending in the plate containing SM elements and for the evaluation of minimum force which is necessary aiming to fix a certain shape of the plate. The numerical research has shown that force which is necessary for the counter-bending of plate varies from 370 to 135 N (it is equal to force being triggered by the plate), when temperature varies from 80 to 100°C.

Numerical results between the maximal displacement of middle point of bended plate and the various reaction stresses are shown in **Figure 5**.

Based on Eq. (6), it is possible to calculate restricted volume V_{PL} of bended plate Eq. (21) and find a changed capacity of exciting chamber:

$$V_{PL}(\sigma_r) = \int_0^b\int_0^l y(\sigma_r, \xi, \zeta)\,d\xi d\zeta \tag{21}$$

$$V_M = V_K - V_{PL}(\sigma_{\max}) \tag{22}$$

$$V_K = \pi {r_K}^2 l_K \tag{23}$$

The biggest volume of exciting chamber V_K when the plate has linear shape is expressed in Eq. (23). By using the above set of chamber and plate parameters, it has been found that volume of exciting chamber could be reduced from $V_K = 1.767 \times 10^{-5}$ m^3 to $V_M = 1.63 \times 10^{-5}$ m^3 (about 7.5%).

The volume of exciting chamber influence on vibrations of working body of vibroexciter is illustrated in **Figure 6**. During the numerical researches [18], after choosing particular geometrical parameters (r_{k1}, l_{k1}), initial tightening Hz, value of mass m_1 of working body of vibroexciter and amplitude of vibrations were $A_1 = 4.51 \times 10^{-3}$ m and frequency $f_1 = 16.81$ Hz (autovibration shape X_1 in **Figure 6**). When the volume of chamber was reduced in 3%, the amplitude of vibrations decreased to a value $A_2 = 4.39 \times 10^{-3}$ m, and frequency increased to $f_2 = 16.92$ Hz. When the volume of camera was reduced in 7.5%, amplitude of

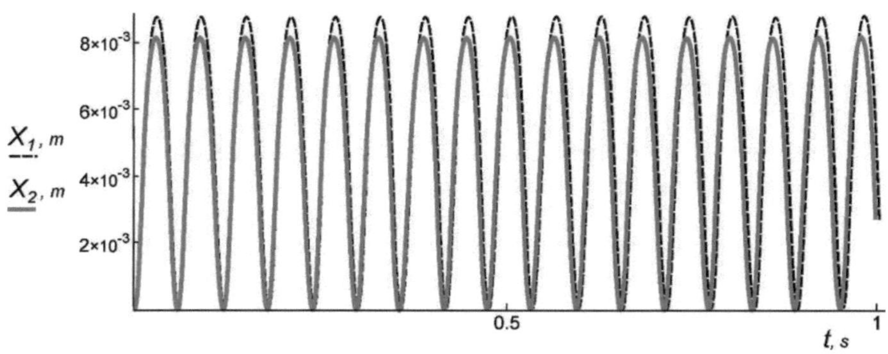

Figure 6.
Autovibrations of mass 1 of vibroexciter in 1 s period: X_1—without element with SMA; X_2—when element with SMA works and decreases the volume of chamber in 7.5%.

vibrations become A_3 = 4.10 × 10^{-3} m and frequency heightened until f_3 = 16.98 Hz (autovibration shape X_2 in **Figure 6**).

2.1 Brinson's model

One dimensional composite mathematical model of internal variables has been offered by Brinson et al. [19, 20].

This model quantitatively evaluates the absorption of deformation energy and allows predicting marginal load which is necessary for the beginning and the termination of a martensitic phase transformation induced by stress [6, 19, 20]. There, martensitic fracture variable is being decomposed into two components depending on stress and temperature.

In this case, the distribution of deformation within the place is not completely linear: material properties vary from martensitic to austenitic state along the thickness of the plate. While bending, there might be three stages of load defined. In the first stage, the SM sample is in austenitic stage along its thickness. In the second stage, when the load is being increased, the martensitic phase transformation, which is being induced by stress, starts in outer layers. Further increase of load provokes plastic deformations that are not included into the energy absorption calculation model [6, 21].

The deformation energy expression according to Brinson is as follows:

$$U = bL\sigma_0\varepsilon_0\left[\frac{h_e}{6} + (h - h_e)\left(\frac{E_2}{2E_1} - 1\right) + \left(1 - \frac{E_2}{E_1}\right)\frac{h^2 - h_e^2}{2h_e} + \frac{E_2}{E_1}\frac{h^3 - h_e^3}{6h_e^2}\right], \quad (24)$$

while martensitic fraction, which is induced by stress, is stated as

$$\xi_S = \frac{1}{2}\cos\left\{\frac{\pi}{\sigma_s^{cr} - \sigma_f^{cr}}\left[\sigma_0\left(1 + \frac{E_2}{E_1}\left(\frac{2y}{h_e} - 1\right)\right)\right] - \sigma_f^{cr} - C_M(T - M_S)\right\} + \frac{1}{2} \quad (25)$$

where σ_0 are ε_0 refer to stress and deformation, respectively, which call for the martensitic transformation; E_1 and E_2 refer to Young's modulus before and after martensitic phase transformation, which had been induced by stress, respectively; σ_s^{cr} and σ_f^{cr} refer to transformation constant; and h_e refers to the thickness of sample core in austenitic stage.

Brinson model evaluates the absorption of deformation energy in SM alloy in both superelasticity and martensitic stages. In addition, it defines the correlation between deformation energy and the thickness of SM plate core.

3. Magnetorheological fluids for use in pneumatic actuators

3.1 Theoretical and experimental background of the micro choke with magnetorheological fluid transmission

Some particularities of the application of magnetorheological (MR) fluid transmissions in microcompressors and pneumatic chokes had been described in scientific works [21–33].

Commonly, magnetorheological fluids are used in three flow modes, as shown in **Figure 7**. Magnetic poles are stationary, and MR fluid is flowing due to a pressure difference perpendicular to magnetic field in valve mode. In the case of shear mode, one of the magnetic poles is moving with respect to the other pole, and the MR fluid

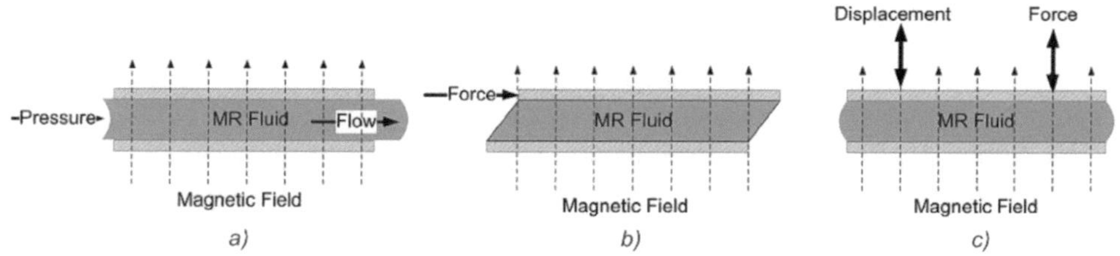

Figure 7.
MR fluid flow modes: (a) valve mode, (b) shear mode, and (c) squeeze mode [31].

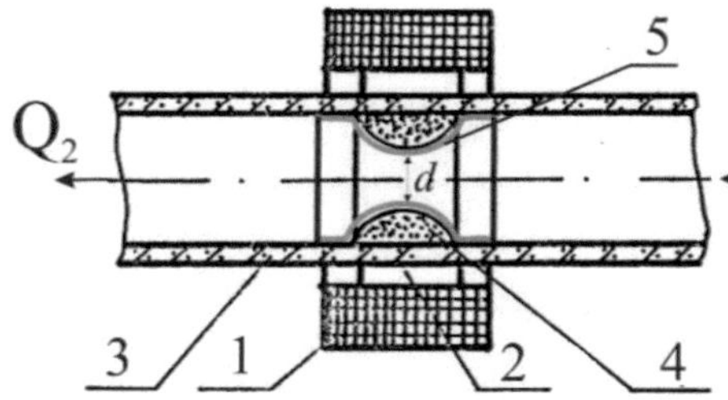

Figure 8.
Micro choke structure containing magnetorheological fluid transmissions: 1—solenoid; 2—constant magnet; 3—tube; 4—magnetic fluid; 5—silicone; and Q_1, Q_2—working medium flow.

is being sheared. MR fluid is placed between two approaching plates and compressed, while magnetic field lines pass through the fluid in squeeze mode [31].

The MR fluid application in valve mode is presented in this research. Magnetorheological effect can be used in controlled choke structures which might be characterised as low power and pressure (see **Figure 8**).

This structure consists of the solenoid *1* with a constant magnet of ring shape *2* and bladdery tube *3* coaxially fitted into its cylindrical opening. Tube *3* is filled with *4*, which consists of the plug of toric segment shape in a silicone bag, along the constant magnet *2*. Under choking, geometrical measures of MR fluid 4 plug and its magnetization force, which is being hold at the internal wall of bladdery tube *3*, depend on the gradient of magnetic field created by the constant magnet *2* and solenoid *1*. If the solenoid *1* is powered by supply voltage from a control unit (no indication of it in the drawing), the field strength and the shape of free surface of MR fluid plug in a working gap change. The diameter *d* of the opening also changes under choking. Thus, it proceeds to the work medium flow regulation.

The substantiation of the mathematical model is based on the ferrohydrostatic equation [33, 34]:

$$gradp = \rho\bar{g} + \mu_0 MgradH \tag{26}$$

and the correlation of pressure capillary jump on a free surface

$$p = p_0 - \alpha(K_1 + K_2) \tag{27}$$

where K_1 and K_2 refer to the main curvatures of normal sections on a free surface, α refers to surface stress coefficient, p_0 refers to pressure of external environment, p refers to fluid pressure, ρ refers to density, g refers to free fall acceleration, μ_0 refers to vacuum magnetic penetration, $M = M(H)$ refers to fluid magnetization and H refers to magnetic field penetration.

First, let us deduce formulas of main curvatures of normal sections. Suppose

$$x = r(s)\cos\phi,\ \ y = r(s)\sin\phi,\ \ z = z(s) \tag{28}$$

It is known that main curvatures of normal sections on surfaces can be expressed through coefficients of the first and the second Gauss form as follows:

$$K_1 + K_2 = \frac{EN - 2FM + GL}{EG - F^2} \tag{29}$$

In this case

$$E = \left(\frac{\partial x}{\partial s}\right)^2 + \left(\frac{\partial y}{\partial s}\right)^2 + \left(\frac{\partial z}{\partial s}\right)^2 = r'^2\cos^2\phi + r'^2\sin^2\phi + z'^2 = r'^2 + z'^2 = 1, \tag{30}$$

$$F = \frac{\partial x}{\partial s}\frac{\partial x}{\partial \phi} + \frac{\partial y}{\partial s}\frac{\partial y}{\partial \phi} + \frac{\partial z}{\partial s}\frac{\partial z}{\partial \phi} = -r'^2\cos^2\phi\cdot r\sin\phi + r'^2\sin^2\phi\cdot r\cos\phi = 0 \tag{31}$$

$$G = \left(\frac{\partial x}{\partial s}\right)^2 + \left(\frac{\partial y}{\partial s}\right)^2 + \left(\frac{\partial z}{\partial s}\right)^2 = r'^2\sin^2\phi + r'^2\cos^2\phi = r'^2 \tag{32}$$

$$L = \frac{\overline{r}''_{S^2}\left(\overline{r}'_{S^x}\overline{r}'_{\phi}\right)}{\sqrt{EG - F^2}} \tag{33}$$

$$N = \frac{\overline{r}''_{\phi^2}\left(\overline{r}'_{S^x}\overline{r}'_{\phi}\right)}{\sqrt{EG - F^2}} \tag{34}$$

where

$$\overline{r}''_{S^2} = \left(r''\cos\phi, r''\sin\phi, z''\right)\ \ \overline{r}'_{S^x} = \left(r'\cos\phi, r'\sin\phi\right) \tag{35}$$

$$\overline{r}'_{\phi} = (-r\sin\phi, r\cos\phi, 0)\ \ \overline{r}''_{\phi^2} = (-r\cos\phi, -r\sin\phi, 0)$$

After integration of Eq. (26) according to z

$$r'' = -z'\left[\frac{\rho g z}{\vartheta} - \frac{\mu_0 M^* H^*}{\vartheta}\log\left(\frac{H^*}{H}sh\frac{H}{H^*}\right) + C - \frac{z'}{r}\right] \tag{36}$$

When s = 0, there are following terms:

$$r(0) = R_0, r'(0) = -\sin\alpha_0, z'(0) = -\cos\alpha_0, \tag{37}$$

And when s = s*,

$$z(s^*) = 0, r'(s^*) = 0, z'(s^*) = -1 \tag{38}$$

The task is solved in the non-asymmetric form:

$$(\overline{rz}')' = \overline{rr}'\left[B_0\frac{\overline{z}}{\overline{r}^2(0)} - A_1\phi(\overline{r}, \overline{z}) + C\right] \tag{39}$$

The digital algorithm with non-inertial variables is investigated:

$$\frac{z_i^{n+1}-z_i^n}{\tau}=\frac{1}{r_i^n}\left(\frac{r_{i-1}^n+r_i^n}{2}z_{s,i}^{n+1}\right)_{s,i}-\frac{B_0}{\left(r_0^n\right)^2}\left[z_i^{n+1}\left(r_{\bar{s},i}^n\right)^+ + z_i^n\left(r_{\bar{s},i}^n\right)^-\right]+r_{\bar{s},i}^n\left[A_1\phi\left(r_i^n,z_i^n\right)-C^n\right] \tag{40}$$

where $i = 1,\ldots, N-1$

$$z_{s,0}^{n+1}=-\frac{h}{2}\left[\frac{B_0}{\left(r_0^n\right)^2}z_i^{n+1}-A_1\phi\left(r_i^n,z_i^n\right)-C^n+\frac{\cos\alpha_0}{r_0^2}\right]\sin\alpha_0-\cos\alpha_0 \tag{41}$$

$$z_N^{n+1}=0;$$

$$\frac{r_{i-1}^n+r_i^n}{\tau}=r_{\bar{s}s,i}^{n+1}+z_{\bar{s}s,i}^{n+1}\left[\frac{B_0}{\left(r_0^n\right)^2}z_i^{n+1}-A_1\phi\left(r_i^n,z_i^n\right)-C^n-\frac{z_{\bar{s},i}^{n+1}}{r_i^n}\right] \tag{42}$$

There are two sets of items $E^n{}_i$ and $F^n{}_i$ such that for each set of elements, C is valid:

$$z_i^{n+1}=E_i^n z_{i+1}^{n+1}+F_i^n \tag{43}$$

After the analytical studies, the relationships are obtained:

$$z_i^{n+1}=\frac{A_i^n}{B_i^n-C_i^nE_{i-1}^n}z_{i+1}^{n+1}+\frac{D_i^n+C_i^nF_{i-1}^n}{B_i^n-C_i^nE_{i-1}^n} \tag{44}$$

and

$$E_i^n<\frac{A_1^n}{B_1^n-\left(1+2c_0h^2\right)C_1^n}<\frac{A_1^n}{A_1^n+1-2c_0h^2C_1}<1 \tag{45}$$

Two sets of ε_i and V_i are determined:

$$r_i^{n+1}=\varepsilon_i r_{i+1}^{n+1}+v_i^n \tag{46}$$

where

$$\varepsilon_i=\frac{\alpha_i}{\beta_i-\gamma_i\varepsilon_{i-1}},\, i>1 \tag{47}$$

$$v_i^n=\frac{\vartheta_i^n+\gamma_i v_{i-1}^n}{\beta_i-\gamma_i\varepsilon_{i-1}},\, i>1 \tag{48}$$

After analytical study is found:

$$r_N^{i+1}=\frac{v_{N-1}^n}{1-\varepsilon_{N-1}} \tag{49}$$

MR fluid magnetic field strength in a working gap has been measured, and the shape of MR fluid plug on free surface has been set [33].

In order to prevent the damage that might be done by pressure transmitted through the micro choke P_1 on the shape of MR fluid plug on free surface under choking, pressure P_1 should not reach the free surface magnetic field force, which has been created.

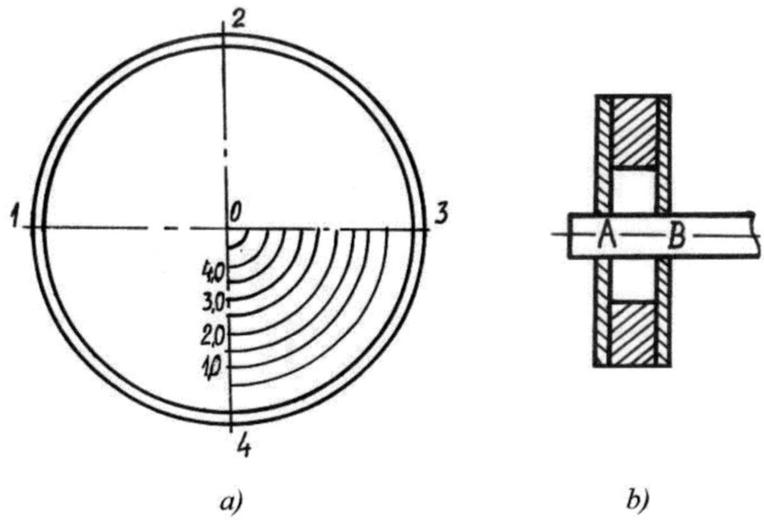

Figure 9.
Scheme of the identifying magnetic field: (a) measuring directions 3–0 and 4–0; (b) measuring points A and B.

Distance from the outside of glass tube h, $\times 10^{-3}$ m	Magnetic field strength H, $\times 10^3$ A/m			
	Direction 4–0		Direction 3–0	
	Point A	Point B	Point A	Point B
0.5	46.871	44.006	47.587	49.179
1.0	36.526	35.651	31.672	34.139
1.5	31.911	24.112	25.942	29.682
2.0	24.987	20.611	19.178	19.417
2.5	20.451	16.870	20.531	15.120
3.0	16.313	10.425	14.801	10.504
3.5	12.812	5.730	11.539	7.639
4.0	11.459	1.751	10.504	4.138
4.5	6.287	1.035	4.218	1.592

Table 2.
MR fluid magnetic field strength in a working gap of the micro choke.

Experiments allowed identifying [33] magnetic field strength at concentrators (points A and B, **Figure 9**) moving from the outside of glass tube to the centre in two perpendicular to each other directions, i.e. from point 4 to point 0 and from point 3 to point 0, in steps of 0.5×10^{-3} m. Research results are presented in **Table 2** and **Figure 10**.

3.2 Study of microcompressor with MR fluid

The operation principle of the microcompressor drive with MR liquid under study has been presented in paper [35]. **Figure 11** shows part of the general microprocessor view.

In the cylindrical system of axes, when the coordinate axis coincides with the rotor axis, the motion of viscous incompressible MR fluid vane is described by Navier-Stokes equations [36], which are expressed as follows:

$$-\frac{\partial p'}{\partial r'}+\frac{1}{Fr_m}\frac{\partial S_m'}{\partial r'}+\frac{R+r}{(R+1)^2}=0 \tag{50}$$

$$-\frac{\partial p'}{\partial \varphi}+\frac{1}{Fr_m}\frac{\partial S_m'}{\partial \varphi}+\frac{\tau\delta(R+r')}{\rho u_0^2}=0 \tag{51}$$

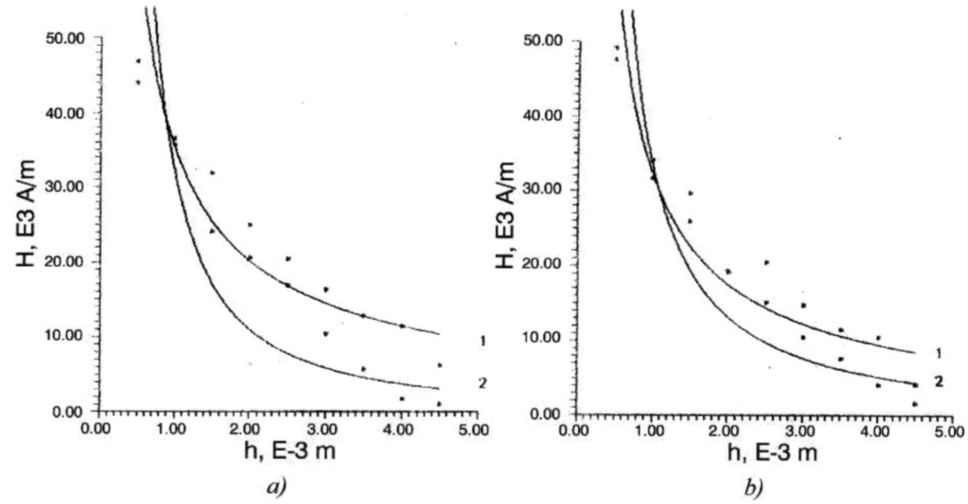

Figure 10.
Correlation between the magnetic field strength H and the distance from a tube wall being measured in two directions: (a) 4–0, 1, measures taken at point A; 2, measures taken at point B; (b) 3–0, 1, measures taken at point A, and 2, measures taken at point B.

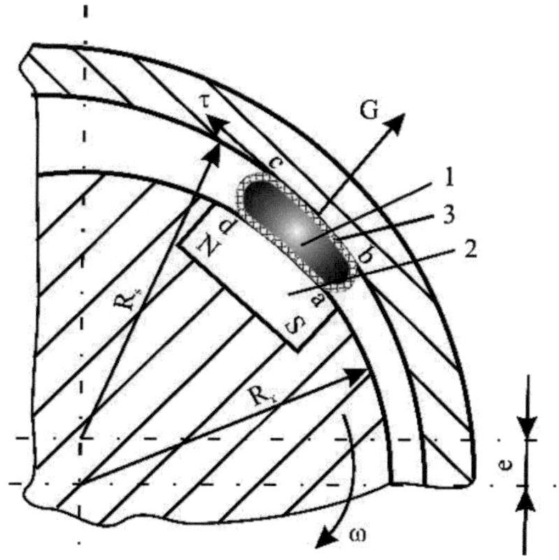

Figure 11.
Scheme of an actuator with MR fluid: 1—MR fluid vane; 2—permanent magnet; 3—silicone shell; R_s*—stator radius;* R_r*, rotor radius; e—eccentricity; a, b and c, d—free surface of MR fluid; G—centrifugal force; ω—rotor angular velocity; S and N—poles of permanent magnet; and τ, shear stresses.*

where $Fr_m = \frac{\rho u_0^2}{\mu_0 M_s H_*}$ is magnetic Froude number; $S_m^{'} = \ln \frac{sh\xi}{\xi}$ is a nondimensional coordinate function of the magnetic field; $r^{'} = \frac{R+r_a}{R+1}$ is a nondimensional relative coordinate; ρ is the density of the magnetic fluid (kg/m^3); δ is the mean thickness of the working aperture between the rotor and the stator (m); u_0 is the vane velocity at the stator (m/s); μ_0 is magnetic permittivity (H/m); $M_s = \varphi_{dal} M_{so}$ is magnetic saturation of the magnetic fluid (A/m); φ_{dal} is the number of particles in a volume unit; M_{so} is magnetic saturation of a particle (A/m); $H_* = \frac{kT}{\mu_0 m}$ is the intensity of the magnetic field (A/m); k is Boltzmann constant; T is temperature (K); $m = V_{dal} M_{so}$ is the magnetic moment of the particle; V_{dal} is the volume of the magnetic particle (m^3); $\xi = \frac{\mu_0 mH}{kT}$ is the argument of Langevin function, $\tau = (R + r^{'})$ $\left[\frac{\partial^2 v}{\partial r^{'2}} + \frac{1}{(R+r^{'})}\frac{\partial v}{\partial r^{'}} - \frac{v}{(R+r^{'})^2}\right]$; $v = u_0 v^{'}$ is the azimuth component of velocity (m/s); and $v^{'}$ is nondimensional velocity.

When studying the efficiency of the drive, it is important to know MR fluid pressure upon the stator surface. When $r' = 1$, the vane pressure upon the stator surface is [35]

$$p^{'} = p_a^{'} + \frac{1}{Fr_m}\left[S_m^{'} - S_m^{'}(0,\ \varphi_a)\right] + \frac{1}{2}\left[\left(\frac{R+r^{'}}{R+1}\right)^2 - \left(\frac{R}{R+1}\right)^2\right] + \\ + \frac{6(R+1)}{\text{Re}}(\varphi - \varphi_a) = \Sigma\left(r^{'},\ \varphi\right) - \Sigma(0,\ \varphi_a) + p_a^{'} \quad (52)$$

The performed calculations have shown that the maximum pressure developed by the actuator reaches 10.8×10^3 Pa, when n = 5000 r/min, q = 0.8 and $\delta = 2 \times 10^{-3}$ m.

For experimental study of an actuator with MR fluid, a stand (**Figure 12**) was made with the following technical parameters: the internal diameter of the stator $D_s = 18 \times 10^{-3}$ m, the external diameter of the rotor $D_r = 15 \times 10^{-3}$ m, the rotor

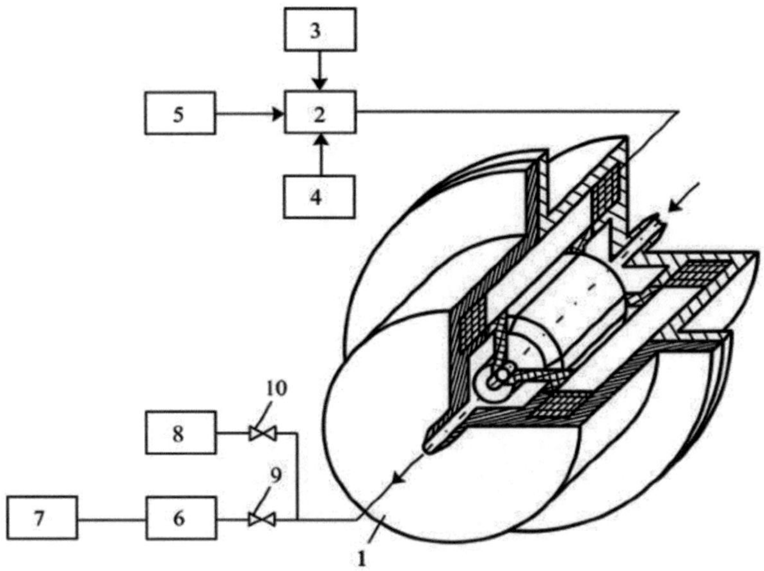

Figure 12.
Scheme of an experimental stand for the actuator with MR liquid: 1—transducer; 2—control panel; 3, 4—power supply; 5—impulse generator; 6—diaphragm; 7—micromanometer; 8—telltale manometer; and 9, 10—valve.

working length $L = 8 \times 10^{-3}$ m, the eccentricity $e = 5 \times 10^{-5}$ m, the volume of the working chamber $V = 6 \times 10^{-7}$ m^3, the number of permanent magnets is 4, the supply voltage $U = 15$ V, the supply current $I = 0.7$ A, the rotation frequency of the rotor $n = 0 \div 5000$ r/min, the MR fluid magnetic saturation $M_s = 51.7$ kA/m, the MR fluid density $\rho = 1585$ kg/m3 and the diameter of magnetic particles $R_{dal} = 9.01 \times 10^{-9}$ m [35].

The distribution of magnetic field in the working aperture of electromagnetic-magnetic systems with MR fluid determines the magnetic-hydrodynamic characteristics of the fluid. Therefore, the magnetic induction in the transducer's working aperture was measured with teslameter (by inserting a sensitive element with Hall sensor into the aperture). The measurements were made both at the surface of the rotor and the stator at 30°. The results of the measurements are presented in **Figure 13**. It has been determined that the highest value of magnetic induction is in the narrowest place of the working aperture at the rotor surface, reaching 0.745 T. The dissemination of magnetic induction at separate poles (magnet) does not exceed 1.2%. The experimental stand was also used for determining the values of the pressure and efficiency generated by the transducer. When the transducer rotor with the magnetic fluid vanes is rotating, the pressure difference developed in diaphragm is registered by micromanometer. According to the manometer readings, the pressure and the efficiency developed by the transducer are calculated.

The dependence of the pressure generated by the transducer on the MR fluid vane-filling coefficient while changing the rotation frequency is shown in **Figure 14**.

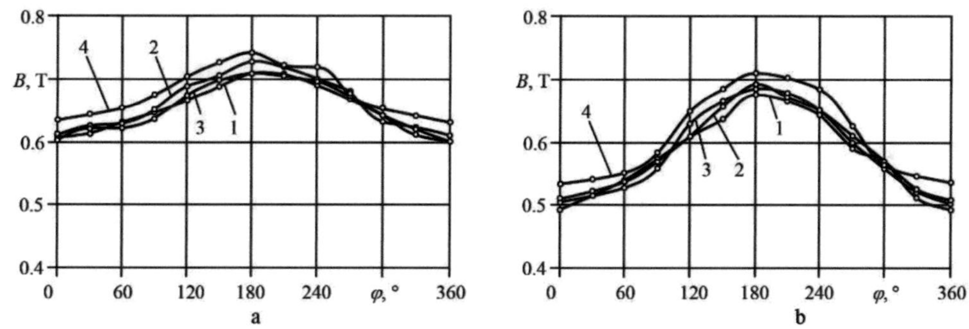

Figure 13.
Dependence of magnetic induction in working aperture upon the rotor turn angle: (a) at the rotor surface; (b) at the stator surface; 1—first magnet; 2—second magnet; 3—third magnet; and 4—fourth magnet.

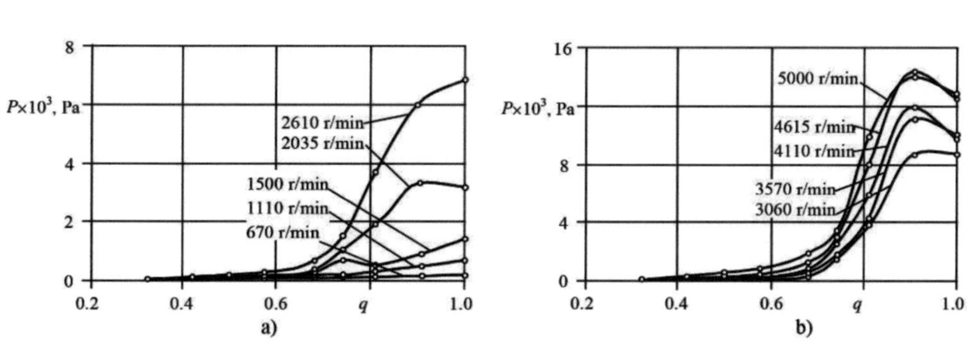

Figure 14.
Dependence of pressure p *generated by the transducer upon the MR fluid vane-filling coefficient while changing the rotor rotation frequency: (a) frequency changed from 670 to 2610 r/min; (b) from 3060 to 5000 r/min.*

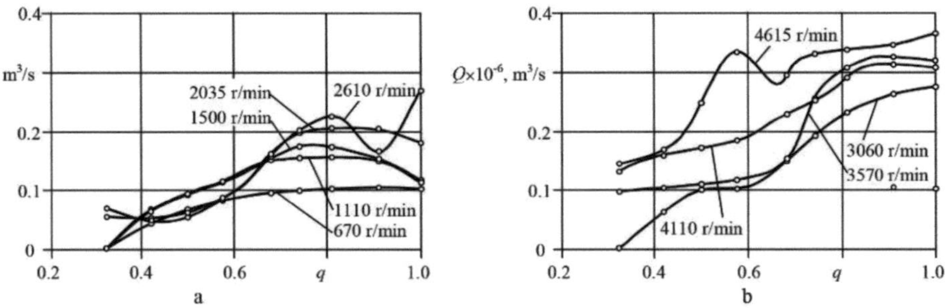

Figure 15.
Dependence of efficiency developed by the transducer upon MR fluid vane-filling coefficient q, when changing the rotor rotation frequency n: (a) frequency changed from 670 to 2610 r/min; (b) from 3060 to 4615 r/min.

As seen from the presented charts, the value of the pressure generated by the actuator depends directly on the MR fluid vane-filling coefficient and the rotation frequency. The highest pressure value is obtained when MR fluid-filling coefficient is close to 0.9. The generated pressure increases with increasing rotation frequency (up to 13.7 kPa, when n = 4615 r/min). **Figure 15** shows the dependence of the transducer's efficiency upon MR fluid vane-filling coefficient while changing the rotation frequency of the rotor.

The study results show that the transducer's capacity also depends on the rotation frequency of the rotor and MR fluid vane-filling coefficient. However, only at low rotation frequencies of the rotor (up to 2000 r/min), the capacity rises gradually with changing MR fluid vane-filling coefficient. The maximum value of the efficiency achieved is 0.37×10^{-6} m³/s.

4. Conclusions

4.1 The performed study of the shape memory elements in pneumatic actuators leads to the following conclusions

The numerical researches have showed dependence of the maximal displacement of middle point of bended plate when both plate's ends are fastened, on the reaction stresses. A numerical methodology has been proposed which describes the changeable volume of chamber with smart material element.

The presented engineering methodology enables to determine dependence of force generated by plate with shape memory material on mechanical stresses.

The theoretical research has shown that the change of chamber volume affects the parameters of autovibrations in vibroexciter. During the studies of the plate of the titanium-nickel alloy TN-1, it was estimated that maximum deformation of plate, whose length $l = 25 \times 10^{-3}$ m, height $h = 1.5 \times 10^{-3}$ m and width $b = 20 \times 10^{-3}$ m, changed the volume of chamber in vibroexciter by 7.5%. It caused an amplitude of vibrations to decrease in 9.1% and frequency to increase in 0.7%. Other SM alloys which have different characteristics and better deformation properties in the direction of Y axis can be used in the construction of vibroexciters. In some technological processes for precise vibroexciters, a minor change of the frequency is also important.

The methodology of calculations used in present research evaluates the deformation of the plate only approximately. For more precise calculations, it is necessary to evaluate the temperature deformations, cross stress and the fact that part of the deformation energy is transferred to the heat used to replace the material structure.

The obtained study results may be applied in designing equipment, containing SMA elements.

4.2 The performed study of the transducer with MR fluid leads to the following conclusions

The value of the pressure generated by the transducer actuator depends directly on MR fluid vane-filling coefficient and rotation frequency. The highest pressure value is obtained when MR fluid-filling coefficient is close to 0.9. The pressure increases with increasing rotation frequency (up to 13.7 kPa, when n = 4615 r/min).

The efficiency developed by the transducer also depends on the rotation frequency of the rotor and on MR fluid vane-filling coefficient. The maximum obtained efficiency value is 0.37×10^{-6} m^3/s.

At low MR fluid vane-filling coefficient q values—up to 0.67—unstable work areas of the transducer caused by hermetic flaws of the system have been observed.

Findings of theoretical and experimental studies are sufficiently adequate; the obtained pressure values do not exceed 15%.

Further studies are needed to ensure hermetic properties of the system with MR fluid vanes.

Thanks

The authors thank Prof. Habil and Dr. Ramutis Bansevičius for the ideas and help in the theoretical research and experiments.

Author details

Edmundas Kibirkštis*, Darius Pauliukaitis and Kęstutis Vaitasius
Faculty of Mechanical Engineering and Design, Kaunas University of Technology, Lithuania

*Address all correspondence to: edmundas.kibirkstis@ktu.lt

References

[1] Kibirkštis E et al. Synchronization of pneumatic vibroexciters on air-cushion operating under self-exciting autovibration regime. Journal of Mechanical Science and Technology. 2017;**31**(9):1-8

[2] Pauliukaitis D, Kibirkštis E. Application of smart materials in structural elements of pneumatic vibroexciters. In: Solid State Phenomena. Zurich, Switzerland: Trans Tech Publications; 2017. pp. 40-44

[3] Stupelis L, Kibirkštis E, Liaudinskas R, Dabkevičius A. Engineering calculation methods of SM element deformation. In: ACTUATOR 2000, 7th International Conference on New Actuators; 19–21 June, 2000; Bremen, Germany; 2000

[4] Ding X, Kuribayashi K, Hashida T. Development of new micro gas valve composed of a SMA thin film and micro distance sensor. In: ACTUATOR 2000, 7th International Conference on New Actuators; 19–21 June, 2000; Bremen, Germany; 2000

[5] Hafez M, Khelfaoui F, Nesnas H, Chaillet N. Monolithic SMA large surface with a high-density of micro actuators for tactile displays. In: ACTUATOR 2004, 9th International Conference on New Actuators; 14–16 June, 2004; Bremen, Germany; 2004

[6] Jia H, Lalande F, Rogers CA. Modeling of strain energy absorption in superelastic shape memory alloys. In: Smart Structures and Materials 1997: Mathematics and Control in Smart Structures. San Diego, California: International Society for Optics and Photonics; 1997. pp. 548-559

[7] Liaudinskas R, Baurienė G, Kibirkštis E, Vaitasius K. Study of Pneumatic Damper with Transducers made from Materials with Shape Memory. Mokslinių tyrimų rinkinys - Mechanika Nr. 3(10). Kaunas, Lithuania: KTU, Technologija; 1997. pp. 46-49

[8] Cisse C, Zaki W, Zineb TB. A review of constitutive models and modeling techniques for shape memory alloys. International Journal of Plasticity. 2016; **76**:244-284

[9] Dutta SC, Majumder R. Shape memory alloy (SMA) as a potential damper in structural vibration control. In: Advances in Manufacturing Engineering and Materials. Cham: Springer; 2019. pp. 485-492

[10] Gao Y et al. An origin of functional fatigue of shape memory alloys. Acta Materialia. 2017;**126**:389-400

[11] Frenzel J et al. On the effect of alloy composition on martensite start temperatures and latent heats in Ni–Ti-based shape memory alloys. Acta Materialia. 2015;**90**:213-231

[12] Savi MA et al. Shape memory alloys. In: Dynamics of Smart Systems and Structures. Cham: Springer; 2016. pp. 155-188

[13] Humbeeck J, Reynnaerts D, Stalmaus R. Shape memory alloys: Functional and Smart. In: 4th International Conference on New Actuatros—ACTUATOR'94, Konferencija; Bremenas, Vokietija; 1994. Pranešimų Medžiaga; Bremen: AXON Technologie Consult GmbH; 1994. pp. 312-316

[14] Hesselbach J, Stork H. Simulation and control of shape memory actuators. In: Proceedings of 5th International Conference on New Actuators – ACTUATOR 96; Konferencija; Bremenas, Vokietija. Pranešimų medžiaga; Bremen: AXON

Technologie Consult Gmbh; 1996. pp. 369-399

[15] Villafana A, Masse M, Portier R, Pons J. Thermomechanical testing machine conceived for the study of shape memory alloys. Application to the training and testing of the two-way memory effect. In: IVth European Symposium on Martensitic Transformations; Journal de Physique IV Colloque; France; 1997. pp. 655-660

[16] Mertmann M, Hornbogen E. Grippers for the micro assembly containing shape memory actuators and sensors. In: IVth European Symposium on Martensitic Transformations; Journal de Physique IV Colloque; France; 1997. pp. 621-626

[17] Patoor E et al. Shape memory alloys. Part I: General properties and modeling of single crystals. Mechanics of Materials. 2006;**38**(5–6):391-429

[18] Pauliukaitis D, Kibirkštis E, Ragulskis K. Problems of precise vibromechanics and vibroengineering. Vibroengineering Procedia. 2016;**8**: 377-385

[19] Bewerse C, Brinson LC, Dunand DC. Microstructure and mechanical properties of as-cast quasibinary NiTi–Nb eutectic alloy. Materials Science and Engineering A. 2015;**627**:360-368

[20] Brinson LC. One-dimensional constitutive behavior of shape memory alloys: Thermomechanical derivation with non-constant material functions and redefined martensite internal variable. Journal of Intelligent Material Systems and Structures. 1993;**4**:229-242

[21] Hao T. Electrorheological Fluids: The Non-Aqueous Suspensions. Amsterdam, The Netherlands: Elsevier; 2011. p. 578. ISBN: 0080455441

[22] Choi SB, Han YM. Magnetorheological Fluid Technology: Applications in Vehicle Systems. Boca Raton: Taylor & Francis; 2012. p. 310. ISBN: 1439856737

[23] Martin Laun H, Kormann C, Willenbacher N. Rheometry on Magnetorheological (MR) Fluids. Rheola Acta: Steinkopff Verlag; 1996. pp. 417-420

[24] Carlson JD. Magnetorheological fluids. In: Schwartz M, editor. Smart Materials. USA: CRC Press/Taylor & Francis Group; 2009. pp. 17.1-17.6

[25] Ghaffari A, Hashemabadi SH, Ashtiani M. A review on the simulation and modeling of magnetorheological fluids. Journal of Intelligent Material Systems and Structures. 2015;**26**(8): 881-904

[26] Fonseca HA et al. Magnetic effect in viscosity of magnetorheological fluids. Journal of Physics: Conference Series. IOP Publishing. 2016;**687**(1):012102

[27] Kordonski W, Gorodkin S. The behavior of a magnetorheological (MR) fluid under compressive deformation. Journal of Rheology. 2016;**60**(1):129-139

[28] Ruiz-López JA et al. Simulations of model magnetorheological fluids in squeeze flow mode. Journal of Rheology. 2017;**61**(5):871-881

[29] Xu F-H et al. A compact experimentally validated model of magnetorheological fluids. Journal of Vibration and Acoustics. 2016;**138**(1): 011017

[30] Rossi A et al. A review on parametric dynamic models of magnetorheological dampers and their characterization methods. Actuators. 2018;**7**(2):16

[31] Farjoud A et al. Nonlinear modeling and testing of magneto-rheological fluids in low shear rate squeezing flows.

Smart Materials and Structures. 2011; **20**(8):085013

[32] Conrad H, Sprecher AF. Characteristics and mechanisms of electrorheological fluids. Journal of Statistical Physics. 1991;**64**(5–6): 1073-1091

[33] Kibirkštis E, Vaitasius K. Investigation of actuators with smart links. Journal of Vibroengineering. 2008;**10**(1):98-103

[34] Stupelis L. An initial boundary-value problem for a system of equations of magnetohydrodynamics. Lithuanian Mathematical Journal. 2000;**40**(2): 176-196

[35] Kibirkštis E. Vaitasius K, Pauliukaitis D. Investigation of actuators with magnetorheological liquid. In: Mechanika 2006: Proceedings of the 11th International Conference; April 6–7, 2006; Lithuania: Kaunas University of Technology/Kaunas University of Technology, Lithuanian Academy of Science, IFTOMM National Committee of Lithuania, Baltic Association of Mechanical Engineering; 2006

[36] Stupelis L. Navier-Stokes Equations in Irregular Domains. Amsterdam, The Netherlands: Springer Science & Business Media; 2013

Chapter 3

Future Prospects: Shape Memory Features in Shape Memory Polymers and Their Corresponding Composites

Safaa N. Saud Al-Humairi, Hasan Sh. Majdi, Amir N. Saud Al-Humairi and Mohammed Al-Maamori

Abstract

Shape memory polymers and their related composites are formally known as SMPs and SMPCs have been classified as innovative categories of smart materials, in which they affected by a particular stimulus and consequently memorize the original shape. As one of the most vital feature of shape memory characteristics, Shape Memory Effect (SME) that been attracted significant attention from the shape memory researchers and scientists. On the other hands, there are abundant approaches can be implemented to actuate the SMPs and SMPCs deformation, whereby the features of the electro-or thermal response associated with the structural changes are predominant. In this chapter, a particular emphasis on how the incorporation of micro/nano-fillers and particles or fibers do affect in the SMP matrices, which it is intentionally carried out to improve the mechanical properties and their related shape memory features of various types of shape memory polymers. In the summary, the shape memory effect is been sustained to be an intrinsic feature for the SMPs and based on this property, the implementation of the SMPs have covered a wide range of applications according to the required functions and performances.

Keywords: shape memory polymer, shape memory effect, shape fixity, thermomechanical cyclic, CNTs, noble metals/fiber-based reinforcement

1. Introduction

As being a kind of smart materials, shape memory polymers (SMPs) and their composites (SMPCs) are significantly attracting consideration [1, 2]. An excellent type of polymers seems to have been disclosed to demonstrate shape memory attributes [2–4], however, the various characterizations or even analysis techniques of shape memory characteristics along with the changing circumstances among various scientists (see **Figure 1**), which they possess intention that the claimed characteristics of SMPs are not identical. Consequently, the correlation between shape memory characteristics and the structures is not entirely recognized for certain categories of SMPs. Thus, this can probably obstruct the growth and

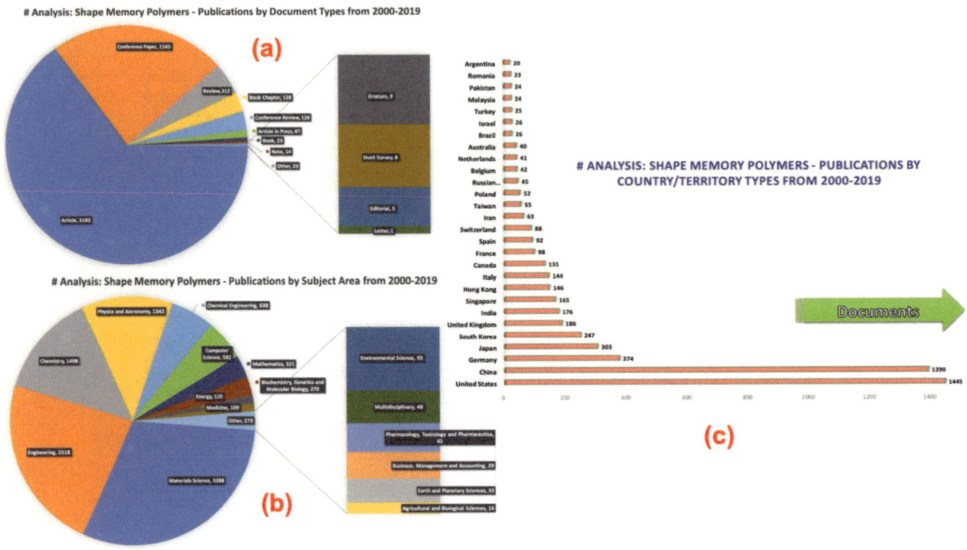

Figure 1.
Shape memory polymer publications based on (a) document type, (b) subject area, and (c) country/territory. "Scopus source accessed on January, 2019".

development of high-performance of SMPs. Aside from that, in comparison with the prompt boost of the variety of SMPs, the main uses of SMPs lags much behind. One among the key factors is certainly utilized the characterization of the SMPs is not going to produce the extensive properties for scientists. For that reason, the analysis of SMPs is vital for the enhancement as well as implantations of SMPs. Based on the Scopus database, a literature analysis was carried out through using the keywords of "Shape Memory Polymers" and/or "SMPs" and the analysis graphs are presented in **Figure 1.** This chapter presents a basic overview up to the date the main employed characterization of the shape memory characteristics of polymers.

2. Shape memory effect in shape memory-alloys over -polymers

The shape memory effect in the shape memory alloys is typically execute based on the test temperature of the austenite ↔ martensite transformation temperature, in which it occurs with the deformation of the SMAs in the martensitic phase during the loading and unloading at temperatures below M_f. After heating these deformed alloys to a temperature above A_f, the austenite phase forms, and thus, the original shape is recovered. In addition, these temperatures are typically will be indicated based on the type of alloys. There are three main based-types of shape memory alloys; Titanium-based, Copper-based and Iron-based SMAs. **Figure 2(a)** shows a typical loading path 1 → 2 → 3 → 4 → 1, wherein the property of SME is observed [5]. The parent phase transforms into the twined martensite (1 → 2) when it undergoes the cooling process. The stress induced detwinning and inelastic strains can occur when the materials are loaded (2 → 3). The maternsite phase is in the same state of the detwinned structure without obtaining any recovered inelastic strains even after the unloaded process (3 → 4). In the final step, the materials are returned to the original shape by recovering the inelastic strains after being heated above Af (4 → 1). On the other hands, there are two types of shape memory effects can be occurred in the SMAs, namely, one-way and two-way SME. On the contrary, the SME of SMPs (see **Figure 2b**) is mainly influenced by the presence of phases

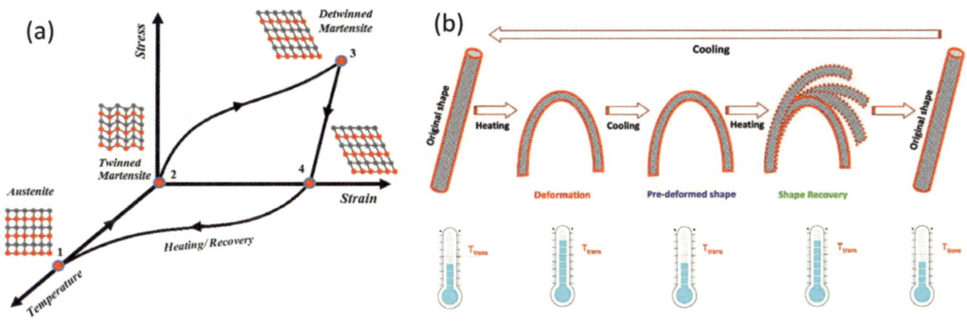

Figure 2.
Schematic diagram of (a) stress-strain-temperature for the involved crystallographic changes during the phenomena of SME [5], (b) one-way SME for SMPs.

Property	SMAs (Ti-based)	SMPs (Polystyrene)
Density/g.cm^{-3}	6–8	0.9–1.1
Deformation strain (%)	<8	≥800
Young Modulus at Temp.> Trans (GPa)	83	0.01–3
Recovery speed (min)	Based on type of alloy	<0.1 min- several min
Deformation stress (MPa)	50–200	1–3
Thermal conductivity (W.m^{-1}.K^{-1})	18	0.15–0.3
Cost ($)	~ 250 per pound	~ 10 per pound

Table 1.
The main comparison between the characteristics of SMAs over SMPs [1, 8].

that linked to the coiled or cross-linked polymer structure. The SMPs is deformed at a temperature below the glass temperature (Tg), and the percentage of deformation is mainly depending on molecular chains of polymer, in which they are controlled by the chemical composition and physical cross-linked structure of SMPs. After preheating the deformed polymers, these molecular chains are able to return back to the original coiled-shape structure. The shape-memory transformation varies according to the apparatus in which polymer molecules transpose between the restricted together with random entangled conformations. As comparison with the SMAs, the SMPs are able to exhibited only one-way SME, whereby, the SMPs deformation at the called phase "soft" only along with the incorporation of the external force [6]. The main benefits of SMPs over SMAs is dependent mainly on their inherent attributes, for instance, they are lower cost and/or density, easier manufacturing process associated with higher percentage of strain [7]. **Table 1** details the principal differences in the SMPs and SMAs characteristics, in which the SMPs are able to obtain up to 800% strain compared with the lower strain in the range of 0.1–20% for SMAs or other types of materials.

3. Shape memory polymer features and properties

To give details about the shape memory characteristics of polymers, a number of variables are essential. Initially, the variables can certainly reveal the characteristics of polymers. Following by the differentiation of them from other sorts of attributes of materials, shape memory capabilities are demonstrated by means of a variety of thermomechanical cyclic procedures. Consequently, the specifications must be able to describe the natural shape memory functions too. Finally, the structure of the variables must look into the prospective purposes. With taken the consideration of these kinds of aspects, a number of variables were presented and also quantified [9–11]. The variables are presented in the following subsections:

3.1 Feature of shape fixity in SMPs

Like appears to have been explained in the foregoing parts, the shape memory behavior is initiated by heating the shape memory polymer to a temperature above the transformation temperature (T_{trans}) [12], it could actually cultivate considerable deformations and this can be mainly predetermined by cooling the materials to a temperature below the T_{trans}, whereby, this parameter was advocated to define the severity of a brief shape becoming fastened in a pattern of shape memorization [12–14]. It ought to be pointed out that the numerous perplexed utilizations as well as illustration appear in the characterization of SMPs. Regarding to the shape fixity, further sorts of capabilities for instance strain fixity [9] and also shape preservation [15, 16], stand for the exact same actual physical indication, in which the shape fixity (R_f) is comparable of the extension ratio of the predetermined deformation to the total deformation, which can be prearranged as:

$$Shape\ fixity = {}^{permanent\ deformation}/_{total\ deformation} \tag{1}$$

Furthermore, the main mechanism of the shape fixity is attributed to the structure and thermomechanical conditions of the shape memory characteristics. The latter condition is significantly been implemented in shape fixity determination as well as the properties of shape memory materials. Wu et al. [17] shows the shape recovery of commercial ether-vinyl acetate copolymer (EVA) with a 300% of pre-stretching at the room temperature, as shown in **Figure 3**. It was revealed that as the number of cycles increased the strain recovery reduced, in which the residual strain starts from 136% and dropped to 112% after 20 min, and within the 9 h, it reaches to

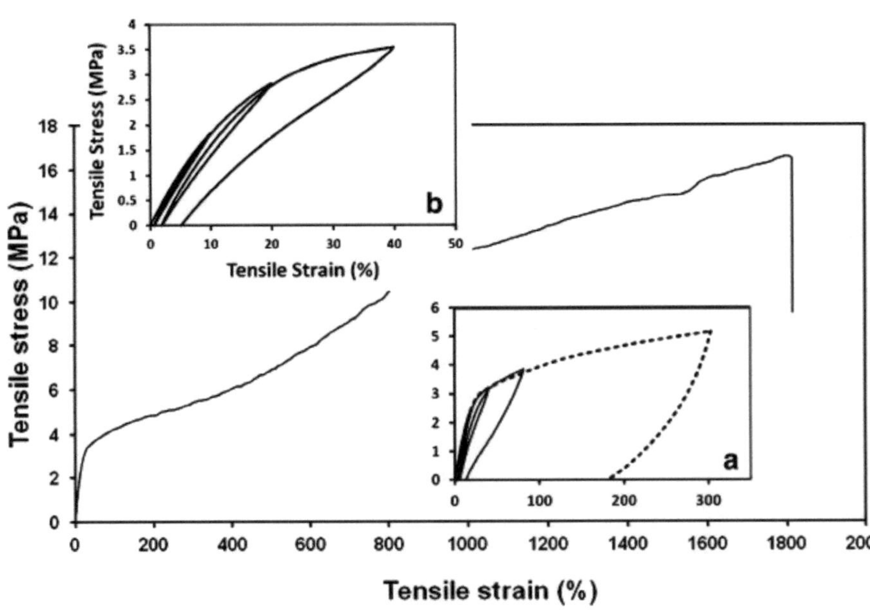

Figure 3.
Relationship curve of stress-strain of EVA at the room temperature. Insets: (a) multiple cycles at different strains; (b) cyclic test [17].

94% and with increasing the time to 72 h, it ends up with 88%. It would be proven that as the time increased; the rapid creep turns to be gradual creep within the first 9 h. Therefore, a long term of the shape fixity ratio was described according to the value of residual strain. Julie et al. [18] demonstrated the shape fixity (R_f) of the epoxy network based on the torsion test, whereby the R_f was found based on the ratio of angle of torsion after unloading to the angle of torsion after loading. It was found that the shape fixity of epoxy within a dimension of $100 \times 10 \times 1\ mm^3$ is about 95% at a deformation angle of 360°.

3.2 Feature of shape recovery in SMPs

Shape memory recovery (R_r) is mainly reflect the ability of any substance to recover the memorized shape after being deformed at low temperature and subsequently heated above the transformation temperature (T_{trans}) [9, 12, 13]. It is significant to notice that, in a shape recovery event, the full strain energy is emitted by means of the two-recovery strain and stress. To a first approximation, nevertheless, the recovery stress to stain ratio is consistent for an identical material [19]. The stored strain energy, alternatively, ought to be influenced by the internal material energy needed in the deformation, despite the fact that the particular relationship is not recognized, in another words, the quantity of energy loss throughout the shape-fixing stage is not identified. Hence, there needs to be a minimum of a qualitative correlation between recovery stress together with deformation energy (or even input energy). Tobushi et al. [9] and Kim and coworkers [13] have performed the thermomechanical test with multi-cycles in purpose of evaluating the performance of SMPs and found the main determination of strain/shape recovery, in which can be calculated using the following interpretation:

$$Shape\ recovery = \frac{Deformation\ recovered\ in\ a\ certain\ cycle}{Total\ deformation\ in\ one\ cycle} \times 100\% \qquad (2)$$

While the shape recovery rate was determined by Li and Larock [20] after been utilized a bending test on SMPs and came with the following formula:

$$Shape\ recovery = \frac{Deformation\ recovered\ within\ heating\ process}{Fixed\ deformation} \times 100\%. \qquad (3)$$

According the above-mentioned equations of (2) and (3), it can be proven that there are different mechanisms referring to the shape recovery in different perspectives. Julie et al. [18] obtained that the kinematic of the shape recovery of epoxy networks is a function of the applied deformation angle. It was found that a complete recovery (100%) was obtained and as the deformation increased the recovery ratio tends to decreased. They also found that the lower heating rate is able to attain a full recovery compared with the high heating rate. As both of two types of recovery are mainly related to the molecular mobility that been coincides with the variation of polymer viscoelasticity properties. They were also proven that the torsion test provides a useful interpretation on the molecular mobility with the glass transition when a uniformed deformation is applied. Moreover, the speed of recovery process and deformation of recovery speed were named by Li et al. [20] and Luo et al. [21], respectively, thereby, both terminologies were reflected the shape memory characteristics of polymers. The shape recovery process of different types of shape memory polymers was studies by Liu et al. [22] using video camera records within a rate of 20 frames per seconds. The results of their experiment revealed that the polymer was capable to obtain a full recovery after 0.7 s. Whilst, Luo et al. [21]

found that the curve of shape recovery of SMPs as function of temperature and then the shape recovery speed was determined based on the following equation:

$$V_r = \frac{dR_r}{dT} \times \frac{dT}{dt} \tag{4}$$

whereas V_r is representing the shape recovery speed, dR/dT is the ratio of shape recovery Vs temperature, and dT/dt is the heating rate. On the other hands, Tobushi et al. [9], Takahashi et al. [23], and Kim [13] were performed the tensile test as thermomechanical cycling via a specially designed machine as shown in **Figure 4a** to study the shape memory characteristics of polymers, in which the tensile test process was isolated under a certain temperature and an extensometer was attached to record the stress–strain data and a programmable software was used to plot the final behavior of one or multicycle of relationship of stress versus strain versus temperature and then the shape recovery and memory effect were determined. **Figure 4b** shows the final curve behavior of shape memory effect, thereby the curve can be classified into four stages, as the first stage, the polymer sample is heated to a higher temperature (i.e. > transition temperature), which often temperature would be in range of 15–25°C. Second stage describes the strain behavior that maintained a constant strain (ε_m) followed with a cooling to a temperature lower than the transition temperature (room temperature) to obtain the permanent shape. The unloading process and the elastic recovery stress turns to reach a zero value at a certain strain (ε_u), as presented in stage 3. An external heating process was applied at a higher temperature to recover the original shape with a minimum value of residual strain (ε_p) as demonstrated in stage 1′ or 4 based on the type of polymer.

Tobushi et al. [21] have demonstrated the shape recovery (R_r) of polymer at different temperatures and the R_r can be calculated based on the following equation:

$$R_r(N) = \left[\frac{\varepsilon_{m-}\varepsilon_p(N)}{\varepsilon_{m-}\varepsilon_p(N-1)} \right] \times 100\% \tag{5}$$

The $R_r(N)$, ε_m, and ε_p are the shape recovery, residual strain, and plastic strain, respectively under a number of cycles (N), in which found that as the number of

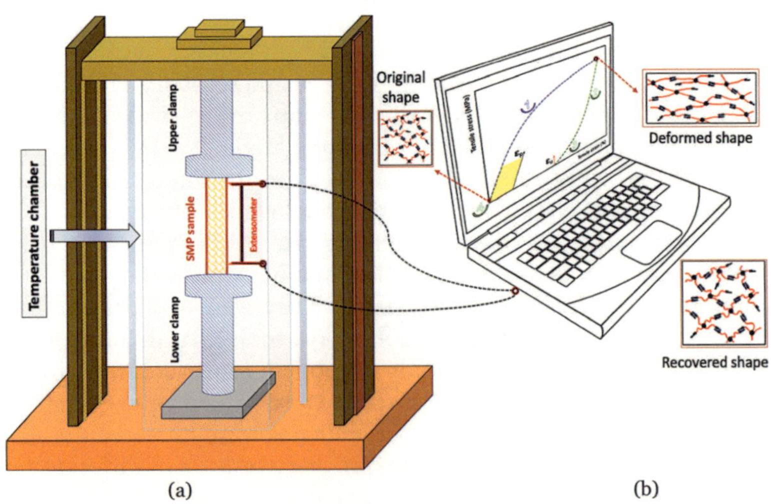

Figure 4.
(a) Tensile test as thermomechanical cyclic machine; (b) shape memory effect curve [9, 13, 23].

cycles increased, the shape recovery maintained to attain 100%. While, Kim and Lee [13] have found that the shape recovery using the following equation:

$$R_r(N) = \left[\frac{\varepsilon_{m-}\varepsilon_p(N)}{\varepsilon_m} \right] \times 100\% \quad (6)$$

Thereby, the shape recovery tends to decrease as the number of cycles increased and shown a stabilized behavior after a number of cycles. Another studies by Liu et al., [22] Lin and Chen [24] and Li and Larock [20] found that the employing of the bending test is much easier and more approachable than tensile test for the thermomechanical test. **Figure 5** shows the mechanism of thermomechanical cycles using the bending test, whereas the shape memory polymer sample with a strip shape is bent to angle θ_{max} at a higher temperature > transition temperature (T_{trans}). The deformed sample followed by cooling process at a temperature < T_{trans}, in which the sample been unloaded and shape recovery started to an angle represented by θ_{fixed}. However, with preheating the deformed sample, the original shape recovered gradually associated with number of θ(T) recorded. At the final stage, the sample turns to recover the final shape at angle of θ_{final}. Therefore, the shape recovery is determined using the following equation:

$$R_r = \left[\frac{\theta_{fixed-}\theta_{final}}{\theta_{fixed}} \right] \times 100\% \quad (7)$$

A compression test was also utilized to obtain the shape memory behavior of polyurethane shape memory polymer foam MF5520 at a nominal glass temperature of 63°C. The shape recovery behavior was obtained as the foams were compressed at T > Tg, then cool it to the room temperature, and finally the shape recovery SMP foam was investigated upon heating after different period time of hibernation. The results revealed that at 80 and 93.4% pre-strain and 1 N applied load, the recovery curves acquired the same trends without a full recovery as shown in **Figure 6a** and **b**, conversely, removing the applied load, the shape recovery was attained. On the other hands, the sample without hibernation displayed a reduction in the shape recovery as the load increased. Gall et al. [25] was investigated the shape recovery of the thermosetting polymer (CTD-DP7 SMP) along with their corresponding composites using a Dynamic Mechanical Analyzer (DMA). The prepared samples were placed in a three-point fixture and the tip of probe was in contacted to the inner surface of SMP

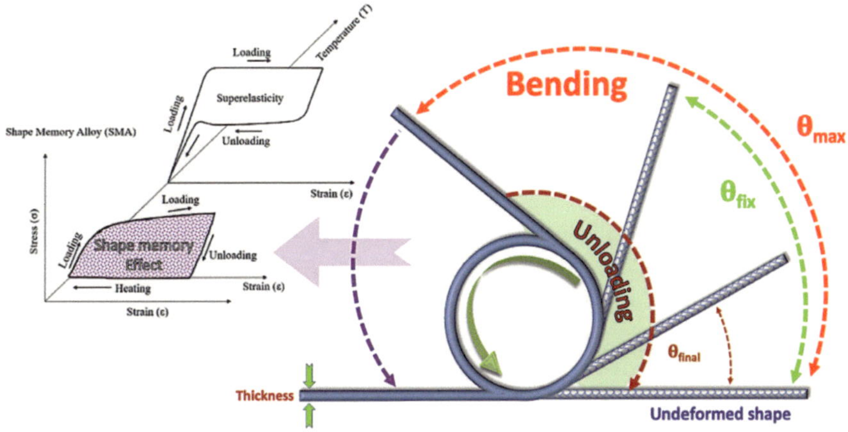

Figure 5.
Schematic drawing of bending test as thermomechanical cycles test.

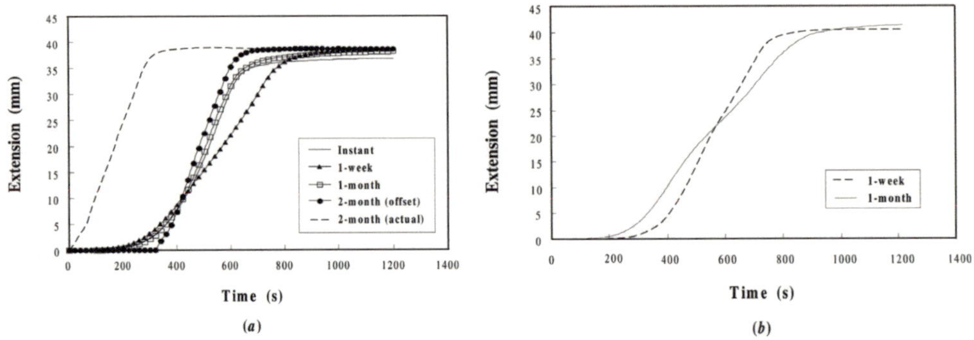

Figure 6.
Extension behavior representing the shape recovery SMP foams hibernated at different time for the pre-strain; (a) 80% and (b) 93.4%.

sample. With fixed probe contact, the shape recovery was recorded as the temperature was increased.

4. Parameters affects the shape memory effect in SMPs

4.1 Thermo-mechanical behavior

Lendlein and Kelch [12] described that the behavior of the shape memory for SMPs is not only linked to the polymer properties, but it is also mainly controlled by the structure and morphology associated with the manufacturing processes. Theretofore, the shape memory effect of any type of polymers is demonstrating the thermomechanical cyclic performance includes, shape recovery, deformation, and shape fixing and each of these processes associated with the condition of thermomechanical process are able to vary the shape memorization and thus affect the shape memory characteristics. Hence, it is essential for further development and advanced applications of SMPs to give a complete characterization. For instance, Wang et al. [26] have carried out the thermomechanical cyclic for SMP composite using the cyclic tensile test within 30 mm/min as displacement rate for 5 cycles, as shown in **Figure 7a** and **b**. They revealed in their results that there is a huge hysteresis between 1st and 2nd cycle, while these differences are attainment smaller from 2nd, 3rd, 4th and 5th cycles. These variations are mainly attributed to the existence of deformation, composite structure failure along with initial training

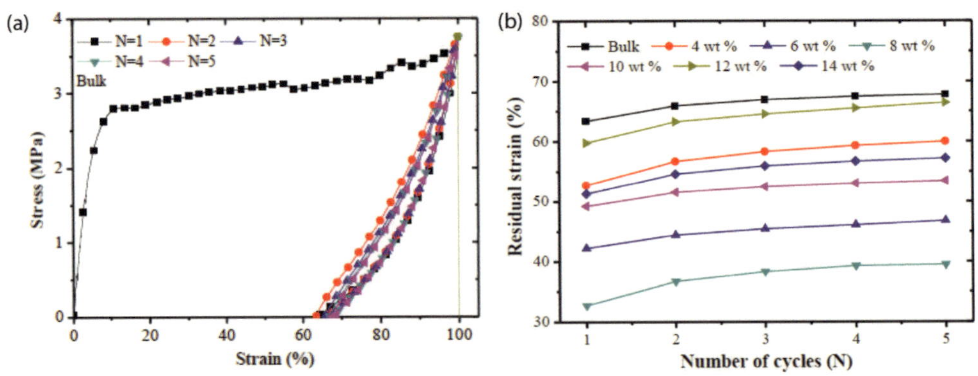

Figure 7.
(a) Stress-strain curves of SMP composite under different loading-unloading cycles; (b) residual strain versus number of cycles of the reinforced SMP with different percentage of chopped carbon fiber [26].

effect for the 1st cycle. **Figure 7b** illustrates that the variation of residual strain at the room temperature versus the number of cycles, in which the residual strain tends to increased precipitously with final stabilization behavior as the number of cycles increased due to the resistance of the modified particles against the deformation, which will be explained in details in Section 3.2.

Pieczyska et al. [27] have been studied the thermomechanical properties of polyurethane theoretically and experimentally under different mechanical loadings at temperature of 20°C above and below T_g at a strain rate of 2 per second within a strain range of 0.6/s. It was observed that when the temperature been slightly dropped, the thermoelastic effect stepped affected. On the other hands, the thermal images (see in **Figure 8**) are referring to the variations in the strain vales during the loading-unloading process, whereby the uniform distribution of the temperature replicated the deformation process in a macroscopically homogenously presented.

4.2 Particle/fiber reinforcement

The design of SMPs and SMPCs thermomechanical behavior can be vary based on the changes in the polymer molecular structure and/or addition of functional particles or fillers in purpose of forming multi-phases composite SMPs. The reinforcement categorization for the SMP composite can be inserted under the particle/filler according to the type of applications. There are various types of particles can be incorporated with SMP, for instance, Silicon carbide (SiC), carbon nanotubes (CNT), nickel, carbon black (CB), clay, and Fe_3O_4 [28–31] along with some others different fibers based on the application requirements [32, 33]. Thus, these types of additions or reinforcements are trigger to enhance the electrical and mechanical properties of SMPs.

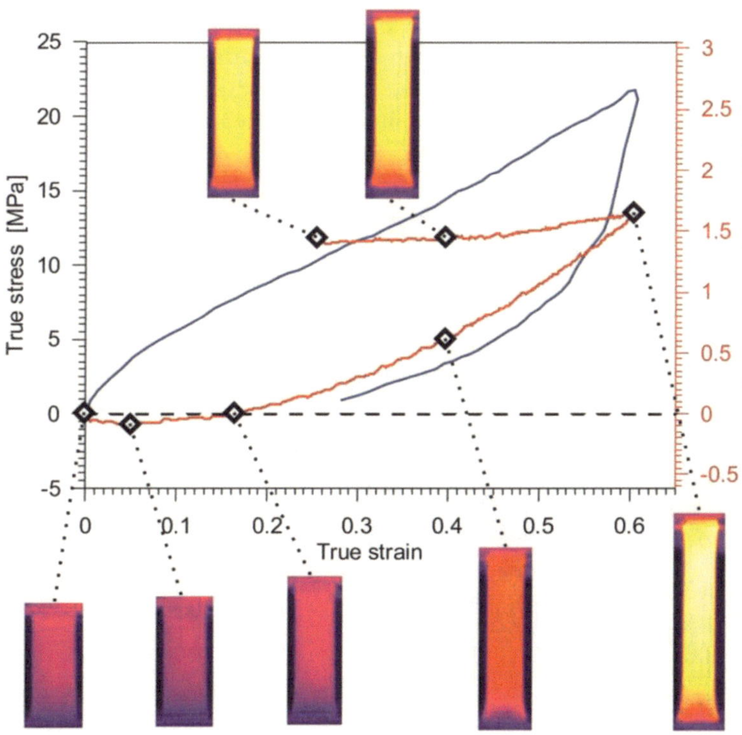

Figure 8.
True stress-strain curve during loading and unloading of PU-SMP along with thermal images that represented different values of strain [27].

4.2.1 Carbon-based reinforcement

Thermally induced shape memory polymers have been characterized as a remarkable substance with a high recovery and shape memory effect is required, however, their mechanical properties such as modulus and strength are still low. Therefore, the incorporation of carbon nanotube (CNTs) with different types of polymer is essential and their inadequate properties are able to improve after a certain modification been considered [8, 34, 35]. On the other hands, the carbon nanotubes have the presented a potentially implementation the nanoelectronics devices, for example, electrochemical energy storage and artificial muscles. According to the above-mentioned deliberations, Kai Yu et al. [36] have studied the effect of carbon nanotubes on the shape memory effect of SMPs after being exposure to microwave radiation. It was found that the CNTs particles have been absorbed the electromagnetic radiation and converted to be an internal heating source, thus lead to induce a shape recovery for SMCs, as shown in **Figure 9**. Furthermore, increasing the frequency of radiation and/or amount of CNTs lead to enhance the shape recovery, whereas the SMPs composite within 3 and 5 wt.% of carbon nanotubes have shown a fully recovery in their shape, while, the SMPs within 1 wt.% addition has a lowered recovery with 80% less than high amount and the unrecovered shape has been resulted due to the insufficient amount of radiation to overcome the caused friction between the CNTs and polymer matrix. From the same point of view, another research was conducted poly(vinyl alcohol) (PVA) filled by CNTs, in which the results revealed that there is a wide boarding in the glass transition temperature and the initiated stress during the recovery was almost double value compared with the conventional polymer [37]. Conversely, Raja et al. [38] found that there no apparent for the shape recovery of PU/PVDF polymer blend nanocomposites after been deformed in "U" shape and preheated using a hot water with a temperature of 60°C for 2 h and followed by a direct quenching in cold water, as demonstrated in **Figure 10a**. An external heating source using a DC controller with a 40 V was attached to the ends of modified polymer strips to activate/initiate the shape recovery. In spite of this, the modified PU/PVDF nanocomposites with CNTs fillers, namely as PUPF-NTM10 has been recovered after 15 s and others samples filled with pristine CNT, namely as PUPF-NTP10 has been recovered the complete shape after 30 s with the external applying of electrical impulse. It was also found that repeating the shape memory test (i.e. increase the number of cycles) led to reduce the shape memory ratio as depicted in **Figure 10b**.

The influence of multi-walls CNTs on the shape memory effect of epoxy nanocomposites was deliberate by Abishera et al. [39] as implementation in the self-healing systems applications under different programming conditions. It was

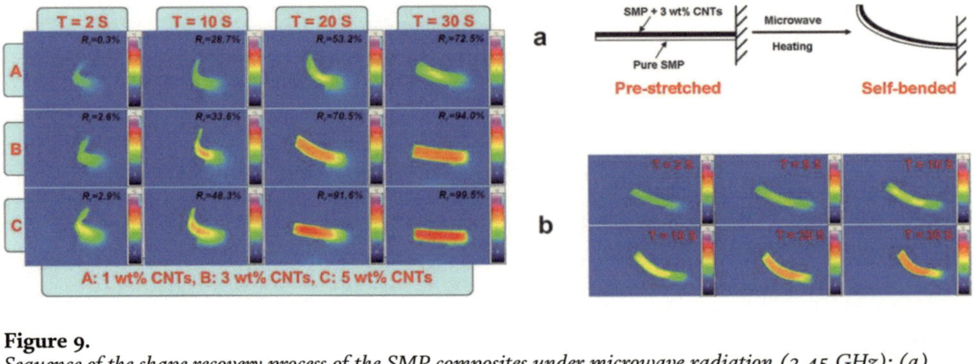

Figure 9.
Sequence of the shape recovery process of the SMP composites under microwave radiation (2.45 GHz): (a) deformation process and (b) modified SMP/CB/CNT sample shape recovery and temperature distributions.

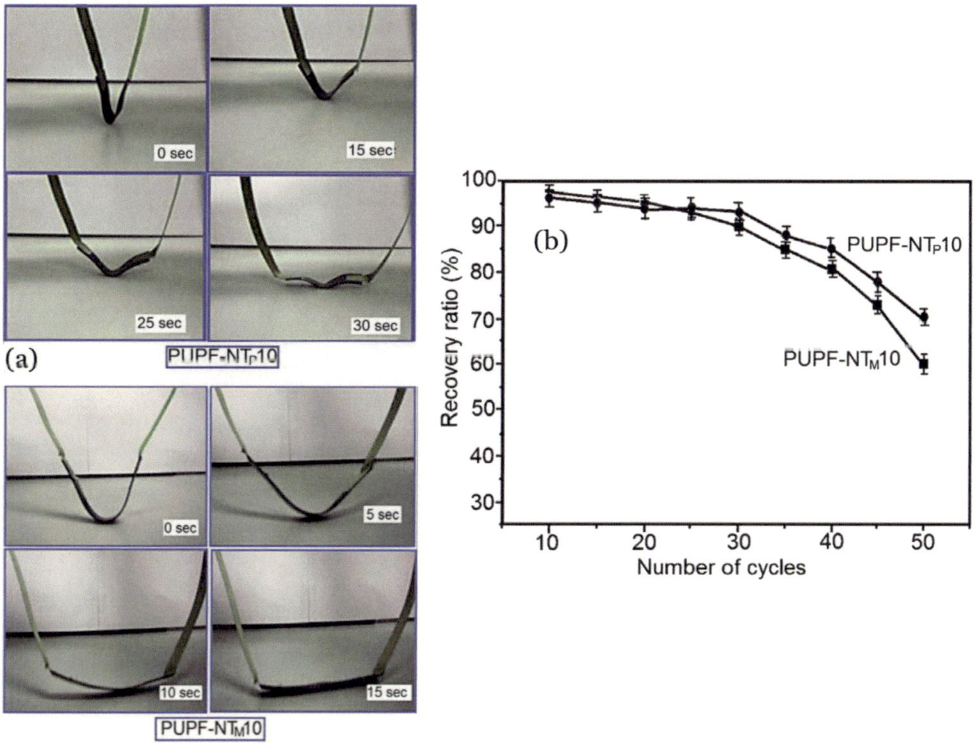

Figure 10.
(a) SME of unmodified PUPF-NTP10 pristine and CNT modified PUPF-NTS10 filled with PU/PVDF nanocomposites; (b) recovery ratio versus number of cycles [38].

revealed that the changing in the programming conditions obtained an excellent shape memory behavior, as well as, the incorporation of the multi-walls of CNTs has indicated an improvement in the young modulus, strength, recovery speed and shape fixity of epoxy associated with drops in the failure strain. Qi et al. [40] have investigated the shape memory properties of polylactide (PLA)/thermoplastic poly (ether)urethane (TPU) composites after been reinforced with carbon black (CB) nanoparticles within the blending ratio of 70:30 by weight. Due to the continuous phase of thermoplastic poly(ether)urethane (TPU), an outstanding shape memory behavior was acquired for the novel ternary structure of PLA70/TPU30/CB, wherein it resulted in the occurrence of the persuasive recovery driving force. In addition, the addition of CB with different percentages displays a slight improvement in the shape fixing ratio (R_f) of 90% (see in **Figure 11**), this may relate to the elongated TPU phase retraction, in which resulted in a total rigid of polylactide phase at 25°C. As the temperature increased more than Tg, the amorphous chains of PLA started to move and thus release constrained TPU phase. The principal contributor in the enhancement of shape recovery ratio (Rr) is the strong pliability of TPU phase and addition of CB, as it was shown only 59% for the binary phase of PLA70/TPU30 and increased to 80.2% as the CB was added, as shown in **Figure 11a** and **b**. Moreover, the increment in the heating time under a consistent 30 V led to increase the shape memory ratio of the ternary phases of PLA70/TPU30/CB6 and PLA70/TPU30/CB8, as shown in **Figure 11c**. The fastest shape recovery response for PLA70/TPU30 after 8 wt.% addition of CB with 90% ratio in 80 s, on the other hands, the PLA70/TPU30 with 6 wt.% of CB approached the same ratio in 150 s, as illustrated in **Figure 11d**. Haibao Lu et al. [41] presented the shape memory behavior of shape memory polymer nanocomposite (SMPs) after carbon nanotube

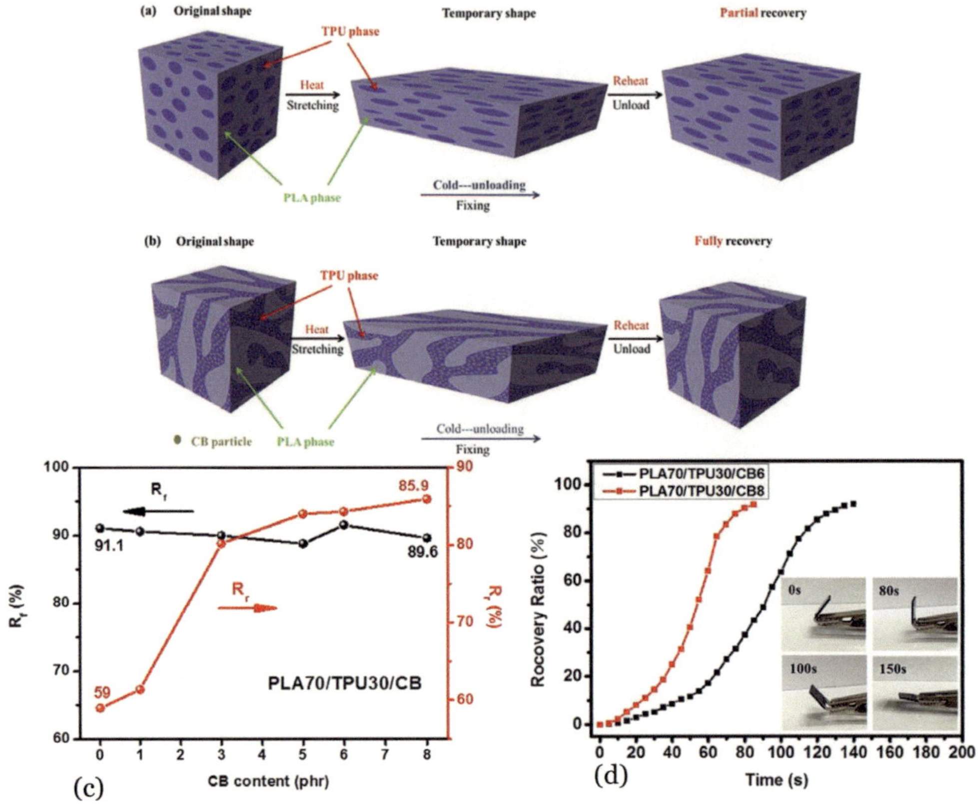

Figure 11.
Representation design of the shape recovery of (a) binary blends of PLA70/TPU30 and (b) ternary blends of PLA70/TPU30/CB5; (c) recovery ratio (R_r) versus CB contents and (d) shape fixing (R_f) versus time PLA70/TPU30/$CB_{6/8}$ [40].

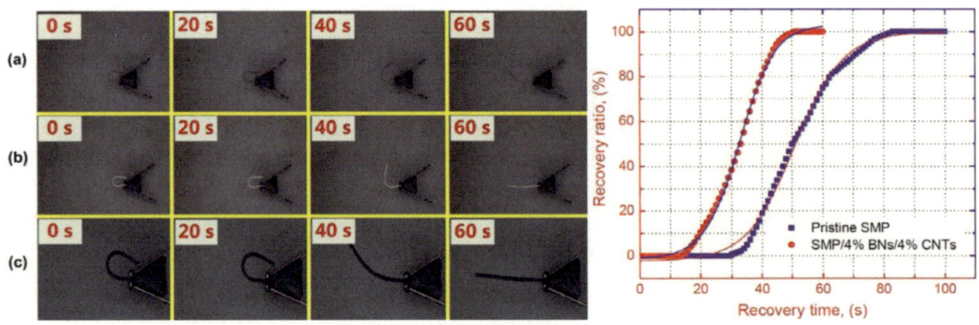

Figure 12.
Optical shape recovery of (a) pristine SMP, (b) nanocomposite SMP reinforced with 4 wt.% of BN, (c) nanocomposite SMP reinforced with 4 wt.% of BN and CNTs, and (d) shape recovery ratio versus time for pristine SMP with and without BN/CNT reinforcements [41].

and boron nitride additions using a bending testing with an Infrared light-induced as heating source. A full recovery 100% in 60 s was recorded for the modified SMP with 4 wt.% boron nitride and CNTs, as shown in **Figure 12(a-c)**, in which the unmodified sample was obtained shape recovery lower than 80% in the same exposure heating time, as shown in **Figure 12d**. This kind of improvements is mainly attributed to the particle of boron nitride that have been improved the thermal conductivity through the facilities of the heat transfer in the composite polymers [44, 45], and thus, CNTs and boron nitride additions have been drastically superior the infrared light-induced shape recovery.

4.2.2 Noble metals-based reinforcement

The effects of the nanosized noble metals-based, such as gold (Au) and silver (Ag) nano-particles/wires on the structure and properties of SMPs have shown a great interest for the researchers and scientists as a multi-responsive shape memory polymer in composite form [8, 42, 43]. Due to the large surface and high plasmonic resonance, these nanosized metals been offered a structure that able to absorb the specific wavelength and convert it into heat energy, thus produce a remarkable type of polymer with high shape actuation and wavelength activation.

4.2.2.1 Ag addition

The cross-link structure between the polymer and ion-metals as ligand coordination have shown a promising method to produce shape memory polymers with noteworthy properties. An isonicotinate-functionalized polyester (PIE) was studied by Wang et al. [44], which they described the effect of silver (Ag) addition to the coordination of the polymer network structure to produce the strip specimens. The shape memory effect was measured based on the DMA test, whereas the strips was heated to 50°C for 1 min, followed by bending process into spiral shape. The shape fixation was carried out by frozen the deformed strips at 0°C for 1 min. Lastly, the strips were recovered the initial shape after being heated to 40°C using hot air. It was found that the modified strip was shown an excellent shape recovery at 37°C for 60 s, as shown in **Figure 13**. Another study on producing a film by Lu et al. [45], the shape memory behavior was investigated using bending test, where the sample was bent in U-like shape at 160°C and cooled to the room temperature (which it was about 22°C). it was found that the sample contains the Ag particles decorated GO has recovered 100% after 36 s within an electric power of 3.87 Watt, as shown in **Figure 14**.

The behavior of the electro-response shape recovery of the surface modified of SMP with Ag nanowires layer was investigated by Luo et al. [46]. The recoded data of shape recovery versus time as shown in **Figure 15a**, the results exhibited that the Ag modified samples were not only able to recover the full initial shape but also with fastest speed recovery in shorten time compared with un-modified samples. It was also found that the higher addition of Ag led to reduce the shape recovery, and therefore, the external heating source was essential. As the heating increase, the thermal transition was occurred and consequently the shape recovery developed. **Figure 15b** shows that with the applied external heat of 5 volt, the bent angle turns to be change within 3 s, this because the presence of nanofillers is capable of conductive network reduction and subsequently increase the resistivity of the strain sensitivity and conductivity [47].

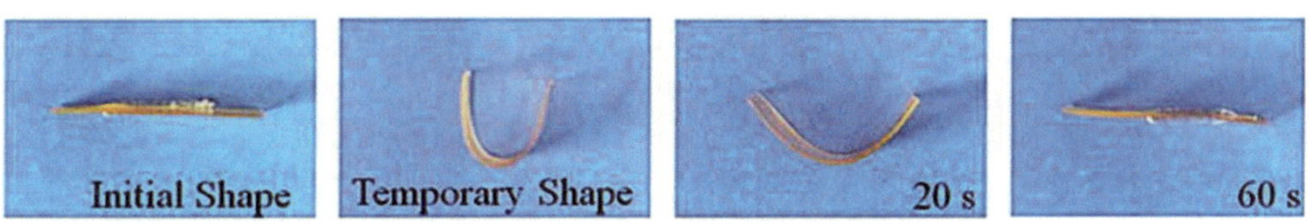

Figure 13.
Shape memory effect of Ag–PIE in PBS at 37°C [44].

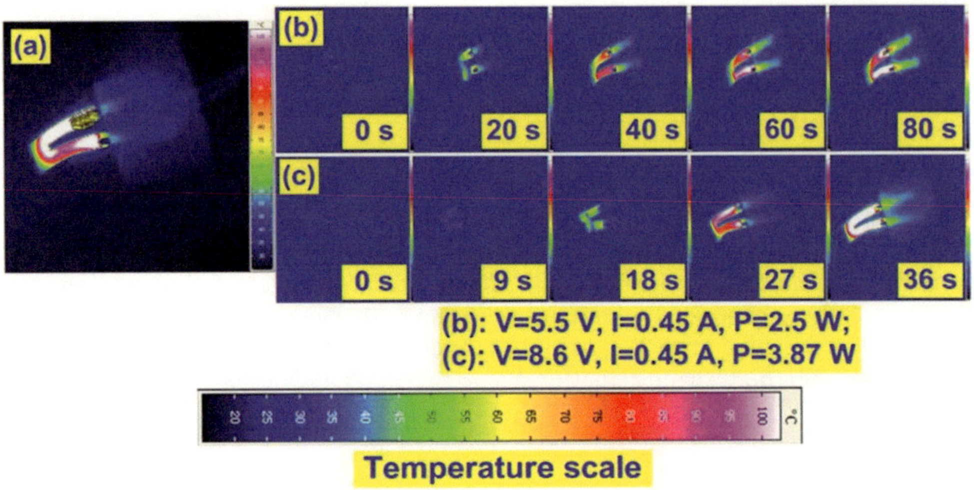

Figure 14.
Shape memory effect of SMPs composite induced by joule heating based on (a) trend scale of temperature, (b) reinforcement of carbon fiber grafted with graphene oxide, (c) reinforcement of carbon fiber grafted with nanoparticles of Ag associated with GO decoration [45].

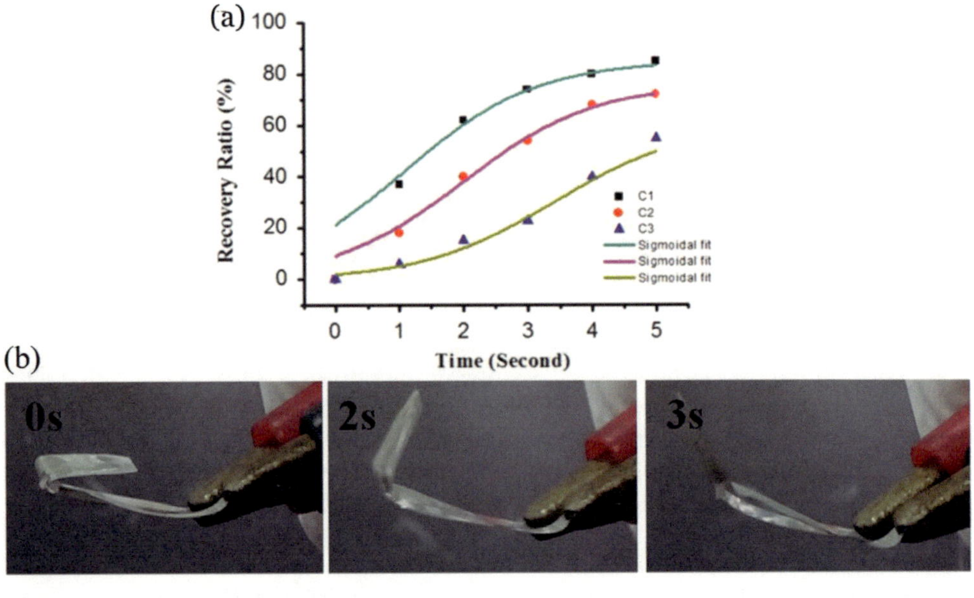

Figure 15.
(a) Shape recovery ratio versus exposure time under a voltage of 5 V; (b) shape recovery demonstration under different exposure times [46].

4.2.2.2 Au addition

A semi-crystalline PEO_{20K} polymer was prepared by Zhang et al. [42] using a cross-linked loaded with nano-particles, whereas found that the gold addition is not utilized to obtain/control the shape memory effect, however, it has also enhanced the properties of self-healing of SMPs. An exceptional concern in such a form of substances is the fact that SMPs necessitate an everlasting system structure (characteristically cross-linked) that could be in confrontation with the substantial a string movability and also inter-diffusion intended for the manufacturing of self-healing properties (SHP) polymers. Within this purpose, the configuration of the

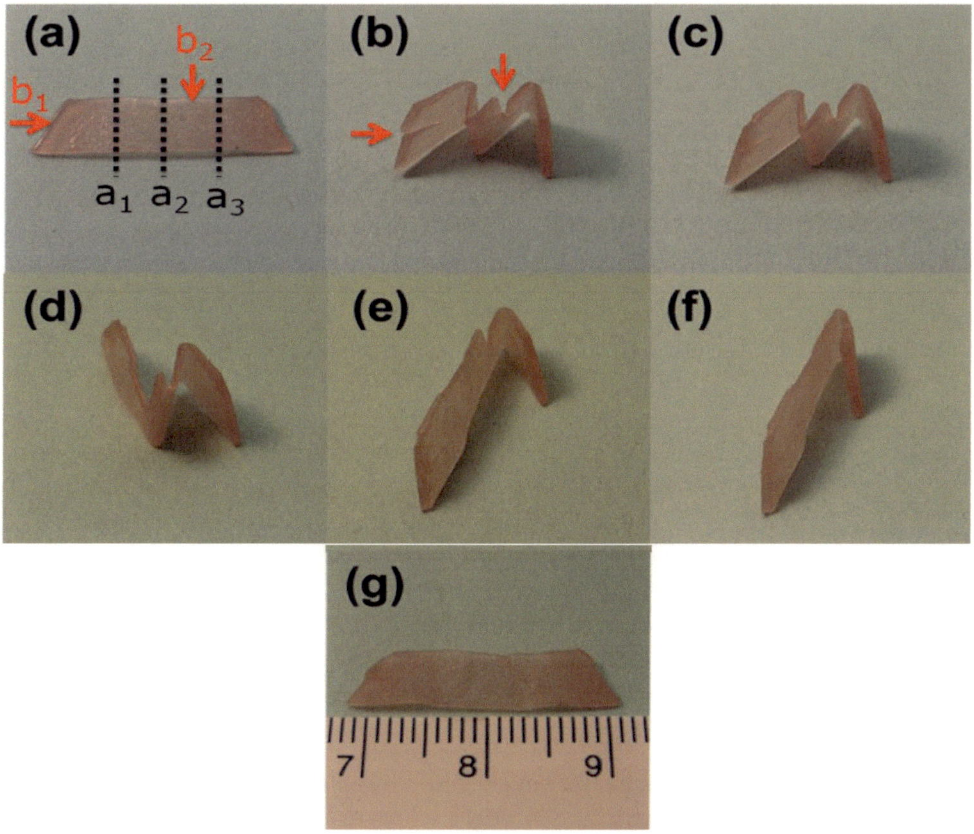

Figure 16.
Optical healing/recovery procedure of a film made from a cross-linked PEO/AuNP using light controlled as a source of heating: (a) Original film, (b) temporary shape obtained by folding the film along the lines a1, a2, a3 at 80 °C followed by cooling to room temperature; then two cuts were made as indicated by red arrows (b1 and b2 in photo a), (c) the b1 cut was healed by exposing the crack to laser (12 W/cm2) for 5 s; (d) the first unbending after 10 seconds laser scanning along the fold a1 at a power of 6 W/cm2, followed by the second unbending under the same condition along the fold a2; (e) the other cut b2 remained in the film of an intermediate temporary shape; (f) the cut b2 was optically healed under the same condition as for the cut b1; (g) the third light-triggered unbending along the fold a3 completed the permanent shape recovery.

SHP it can be more suitable to develop materials upon one single-polymer with the two light-controlled shape-memory and optical recovery features. It was found that the addition of small amount of AuNPs of 0.003 wt.% to the cross-linked poly (ethylene oxide) (PEO) films was sufficient enough to deliberate the main shape memory properties of produced films. The shape memory effect of the produced films is shown in **Figure 16**.

In 2009, Hribar et al. [48] have been developed a modified polymer network, namely β-amino esters, whereas the modified nanocomposites SMPs has demonstrated shape memory effect using IR light as heating source within a temperature above glass transition temperature (Tg). The results exhibited that the prompt shape recovery was spotted until the transformation turn out the path of the beam associated with a thermal transition in the polymer from glassy ⇒ rubbery networks. There are new perspectives for the actuated and functional shape memory polymer that recently been used in different application, for examples: dry adhesives, panels of light-tracking solar, light-guided for smart windows and even actuators. Zheng et al. [49] have been produced a light responsive SMPs as micropillar with a diameter of 10 μm mixed with 0.1–0.2 mol% of AuNRs in a hexagonal array

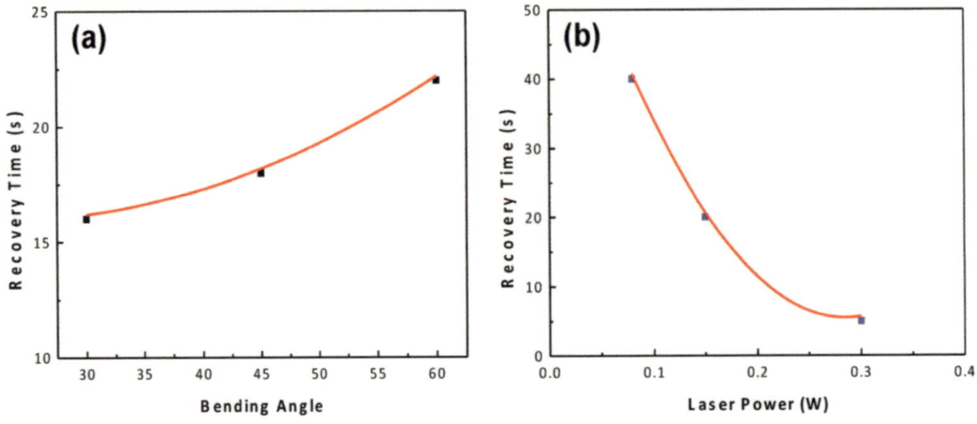

Figure 17.
Recovery time versus (a) bending angle and (b) laser power, of the reinforced SMPs with AuNRs at 45° pillar angle [49].

using poly(dimethylsiloxane) mold as replica molding. To obtain a temporary shape, the pillars were bent at different angles (θ = 30, 45, and 60°) at a temperature above the glass transition temperature (i.e., T = 20°C) and then following by cooling process to the room temperature, as shown in **Figure 17a** and **b**. A heating source was applied using a green laser with a wave-length of 532 nm for different period of times and the shape recovery was monitored by optical microscopy. The results revealed that within 0.08 W, as a laser power, the pillars need 40 s to recovery their temporary shape, however, with raising the laser power to 0.3 W, they grabbed the original shape within only 5 s. It was also found that when the pillars are ground to be collapsed, the shape recovery will not be able to obtain a full recovery (i.e.,100%), and this may attribute to the large force of adhesion between the substrate and pillars, in which it has particularly a higher value than the stored elastic energy of the deformed pillars.

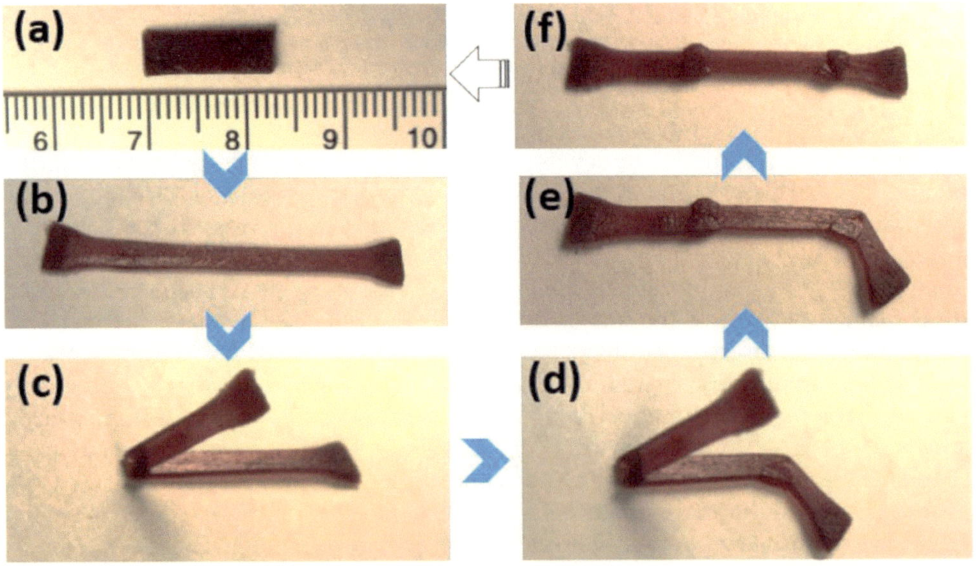

Figure 18.
Optical observation of the shape transformation of cross-linked PEO film reinforced with AuNPs using a laser as a temperature source; (a) original film, (b) deformed sample at 80 °C, (c) large-out of plane bending, (d) gradient chain relaxation and (e,f) shape recovery after applying the laser heating source.

A cross-linked poly(ethylene oxide) (PEO) loaded with 0.5 wt.% AuNPs was experienced to bending loading and unloading along with an exposure to a laser source within a wave-length of 532 nm and power of 0.15 W [50]. The deformation process involved a stretching of the modified PEO/AuNPs to 90% (approximately 200% strain) at 80°C above the Tm, followed by cooling to the room temperature. The top surface of the film was exposure to the laser in purpose of maintaining the anisotropic relaxation in the polymer chains, followed by cooling a temperature below Tm and consequently, resulted in fixing the temporary shape, as shown in **Figure 18(a–f)**, in which the main principles of the technology to exploited the unique shape memory effect property in numerous types of photothermal based shape memory polymers. On the other hands, the adaption of the light polarization has also shown a significant effect on the shape memory property, in which it led to control the photo-based thermal effect, as shown in 2013 study by Zhang et al. [51], whereas the cross-linked network polymer of PVA containing of 0.02 wt.% AuNRs film was stretched and heated to a temperature of 80°C, i.e., above the T_g to maintain the shape permanent transformation, as shown in **Figure 19(a–c)**. A laser in a linear polarization featured with a wave-length of 785 nm and 0.2 W/cm^2 was applied. The results revealed that there is no shape recovery was obtained when the polarization was perpendicular to the deformed film, even after 2 minutes of exposure. Whilst, the deformed shape was fully recovered in 10 s when the polarization turned to be in parallel direction aligned with stretching direction. The reason behind the directional effects can be elucidated that there was no longitudinal absorption in the perpendicular direction, thus resulted with no heat released. However, with the parallel direction, the longitudinal absorption reached to the maximum, and subsequently, the temperature increased above the glass transition temperature (T_g) that lead to reactivate the shape recovery.

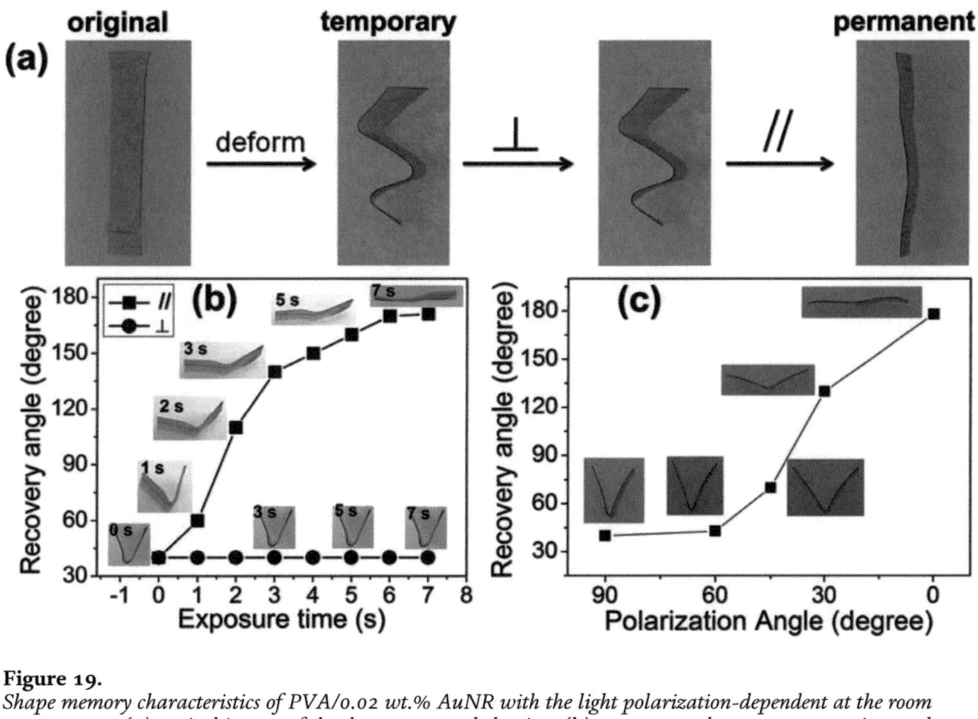

Figure 19.
Shape memory characteristics of PVA/0.02 wt.% AuNR with the light polarization-dependent at the room temperature; (a) optical images of the shape memory behavior; (b) recovery angle verse exposure time under different period of times in two directions, parallel and perpendicular; (c) relationship curve of recovery angle and polarization angle in 1 minute exposure using a light polarization [51].

4.2.3 Fiber reinforcement

Generally, the enhancement in the shape memory polymer mechanical properties is still limited and their reinforcement with short fibers and/or particles is irrationally to be proposed as structural substances [52]. Therefore, the reinforcement with continuous fiber has been attentionally employed to improve the mechanical properties of SMPs [53, 54]. Due to the highly potentialized SMPs, their usages are widely fulfilled for various advanced applications, such as solar arrays, trusses and antennas, which they essential with no moving segments [32]. Furthermore, the majority of researches with regards to SMPs composites are involved in thermoplastic SMPs resins such as polyurethane SMPs. Nevertheless, the comparatively inadequate thermal together with mechanical properties, for instance, moisture, temperature and/or chemical resistance of thermoplastic SMPs are not able to fulfill sensible demands [2]. Thermosetting SMPs, in spite of this, established a marked improvement in the latter characteristics which enable it to be extremely popular for many practical and or structural resources. The development of the fabricated polyurethane SMPs reinforced with carbon-based fiber was implemented for the industrial applications [55–57], a larger ratio of the bending recovery was exhibited in the reinforced polymers compared with pure SMPs sheets, from the same point of view, the epoxy-based SMCs namely EMC (Elastic Memory Composite) was potentially executed for the structure of spacecraft applications, as this EMC was developed in the early 1990s by Composite Technology Development (CTD) [58, 59]. Gall et al. [60] have been investigated the deformation micro-mechanisms of EMC and highlighted the development interaction reinforcement between the epoxy EMC laminate as shape memory polymers resins and fibers and found that because of the changing in the surface of the neutral-strain and micro-buckling effects, the reinforced SMP was able to produce a large value of compression strain compared with the traditional resin composite. Furthermore, the development of composite of thermosetting styrene-based reinforced with fiber was studied by Leng et al. [61], and found due to the good strain capabilities and their relative properties, these types of reinforced-polymers have been potentially chosen to take a part of the structure's applications. From the same perspectives, as a comparison with the pure SMPs, the carbon-based fibers present better thermomechanical properties, and thus been proposed to be use as multi-functional materials [32, 62]. The reinforcement of the glass and kevlar fibers SMPs composite have obtained a superior improvement in the stiffness associated with decrement in the recoverable strain, has been premeditated by Liang et al. [52]. On the other hands, the chopped glass fibers were added to the thermoplastic SMPs and their influences on the shape memory characteristics were examined by Ohki et al. [55]. It was found that the reinforcement by 50 wt.% glass-fiber entertains increment in the failure stress to 140% and reduced the recovery rate to 62%. Another study by Wang et al. [26] demonstrated the shape memory effect of TPI SMPCs under the effect of different mass fraction of the chopped carbon-fibers at three experimental temperatures of 299, 319, and 339 k, as shown in **Figure 20a**. it was found that both experimental variables; temperatures and fibers reinforcement have been remarkable affected the shape memory effect of TPI polymers. The shape recovery ratios tend to rapidly increased as the applied temperatures increased and reinforcement ratio of carbon fiber decreased (see **Figure 20b**), which can be justified these variations that as the cross-linked structure are getting dense, their movements are required more energy and the chain segment movements would be required a higher free space. Therefore, as the temperatures increased, the chain segments will have enough energy to release and thus perform the shape recovery.

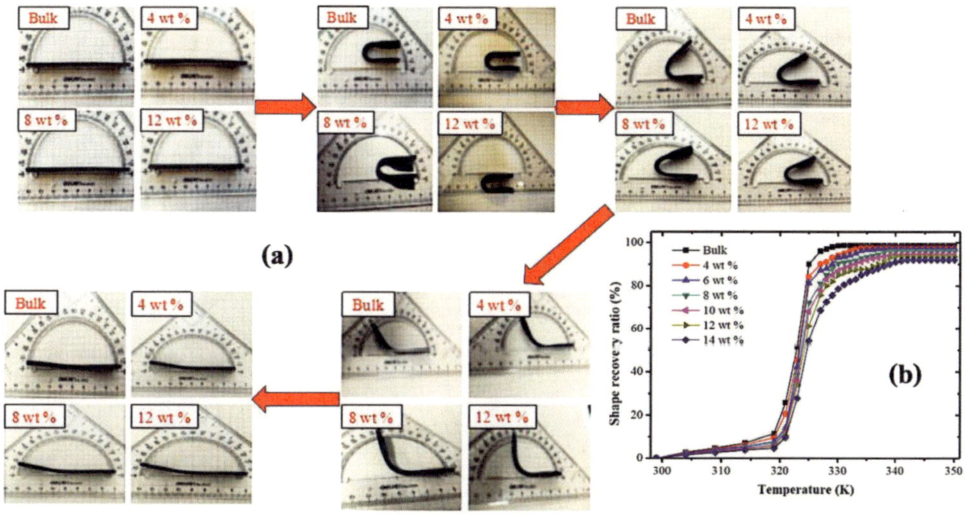

Figure 20.
(a) Shape memory effect behavior with the reinforced SMP along with (b) the shape recovery ratio [26].

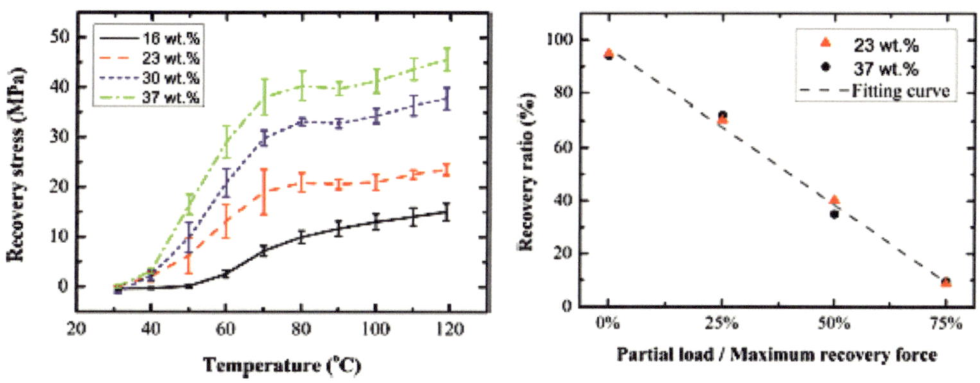

Figure 21.
(a) Stress recovery curves for four percentages of fiber reinforcement; (b) recovery ratio of the reinforced SMP composites with 23 and 37% mass fraction within different partial loads [63].

Figure 21a and **b** displays the epoxy-based SMPs reinforced of carbon fibers with four percentages ratio of 16, 23, 30, and 37% were designed by Li et al. [63]. The designed composites have demonstrated an excellent shape recovery at 120°C within a ratio of 90% and obtained a full recovery 100% after 20 min at the same temperature. The recovery ratio behaved proportionally with the mass fraction of fibers and inverse proportional with the partial load level. Extensive researches were carried out on different types of SMPs matrix and various fiber reinforcements because of the increasing demands for different applications as been summaries in **Table 2**.

5. Conclusions and remarkable observes

Shape memory polymers (SMPs) and their composite (SMPCs) have exhibited exceptional features that led to proposed them to be implemented as advanced materials for the current and potential applications. However, the traditional shape

Author	Substrate	Type of fiber	Recovery ratio (%)	Shape fixity ratio (%)	Comments	Shape memory recovery method/demonstration
Alvaro Iregui et al. [64]	Polymerized epoxy/ polycaprolactone blends	Blends	88–100	95–99	• A remarkable shape memory effect was obtained from the designed mat of the reinforced SMPs.	 (c-e) Shape recovery mechanisms for three times folding at 60°C for 30 min
Paul Robinson et al. [65]	Thermoplastic Interleaf Layers	Carbon Fiber Laminates	fully	—	• The recovery times were significantly longer for the quasi-isotropic laminates (8–9 min) than for the cross-ply laminates (2–3 min).	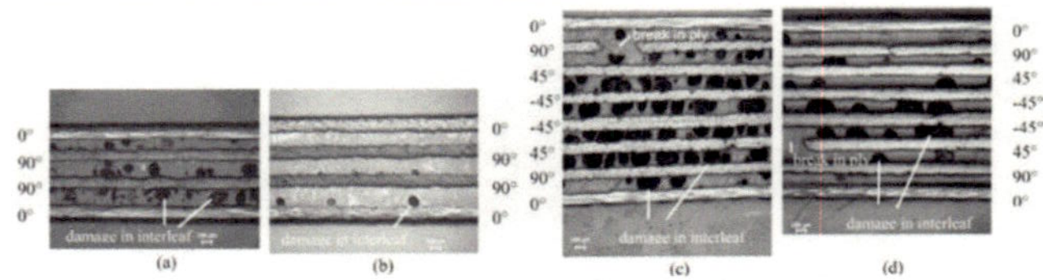 (a-d) Micrographs of the cross-sections after the shape memory recovery process under different bending angles
Jianming Guo et al. [66]	Trans-1, 4-Polyisoprene (TPI)	Carbon Fiber	30–70	—	• The shape recovery ratio decreases as the carbon-fibers fractions increases. • The temperature effects have shown a significant enhancement, as the temperature increased, the shape recovery increased.	Bend and fold method
Guoqiang Li [67]	EPOXY EPON828	Polyurethane	24	30–66	• The stress–strain recovery decreases nonlinearly. • The applied force to close the initiated crack in the matrix led to narrow down the area of the cracks.	Three-point bending test

DOI: http://dx.doi.org/10.5772/intechopen.84924

Author	Substrate	Type of fiber	Recovery ratio (%)	Shape fixity ratio (%)	Comments	Shape memory recovery method/demonstration
Guoqiang Li [68]	EPON™ Resin 828 (thermosetting polymer)	Polyurethane	—..	—..	• The relationship between the healing efficiency and pre-strain level is indirectly proportional. • The addition of 50% fibers caused an improvement in healing efficiency to double value compared with nonmodified SMPs. While, further pre-straining did not show any significant improvement due to the saturation behavior of the cracks in the healing efficiency.	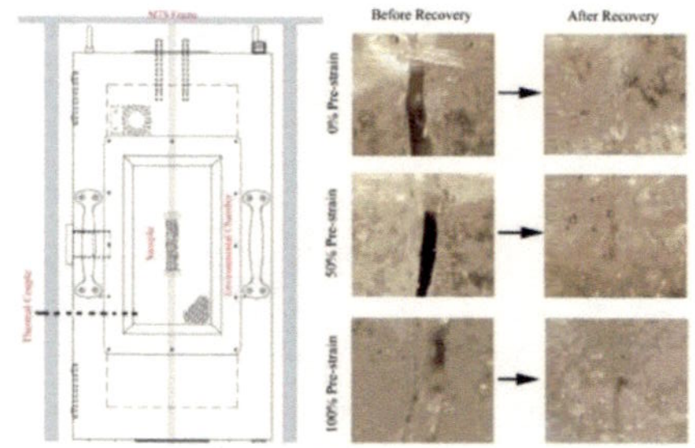 Schematic and optical images of the healing test and shape recovery
Herath et al. [69]	Epoxy Prepreg	Carbon Fiber	86	100	• An excellent shape memory characteristic attained in the modified shape memory polymer composite with carbon fibers without any existence for the effects of spring-back.	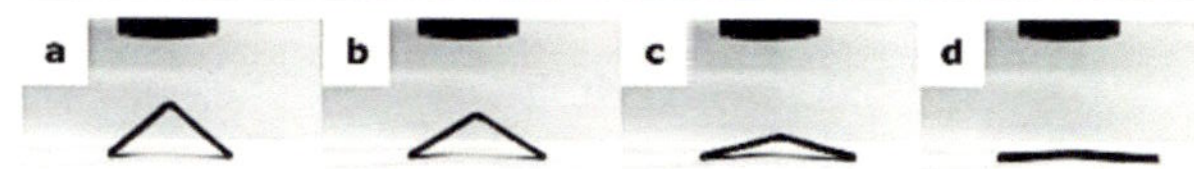 Steps of the shape recovery for the modified SMPCs at different bending angles.

Author	Substrate	Type of fiber	Recovery ratio (%)	Shape fixity ratio (%)	Comments	Shape memory recovery method/demonstration
Qiao Tan et al. [70]	Epoxy Based	T700 Toray Carbon Fiber	—	—	• The glass ↔ rubbery transition concepts have presence in harmony behavior as the temperatures and amount of fiber changed.	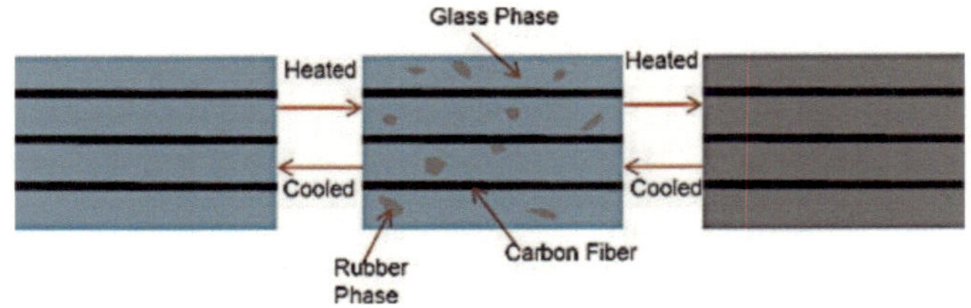 Glass ↔ rubbery transition state for the SMPCs/CF.
Shankar A.Hallad [71]	Epoxy	Nano- and Micro-Graphene and Carbon Fibers	—	—	• The SMP beams have revealed an enhancement in the mechanical properties as the reinforcement with GO and CF existed comparing with the nonmodified SMP beams.	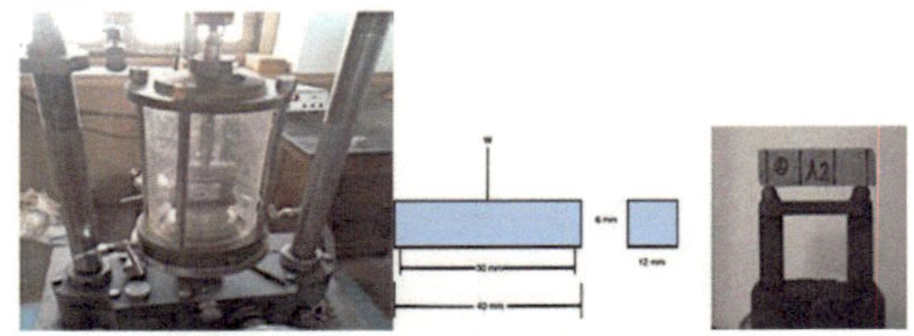 Three-Point Load Set-up for Shape Recovery Test
Haibao Lu [72]	Epoxy-Based Thermoset SMP Resin	Carbon-Fibers	95%.	—..	• The combination of GO and CF additions have been significantly the reliabilities during the bonding procedure associated with assisted the heat transfer resistivity between the matrix of polymer and fibers. • A synergistic effect of rGO and CF on the induced shape recovery during the Joule heating.	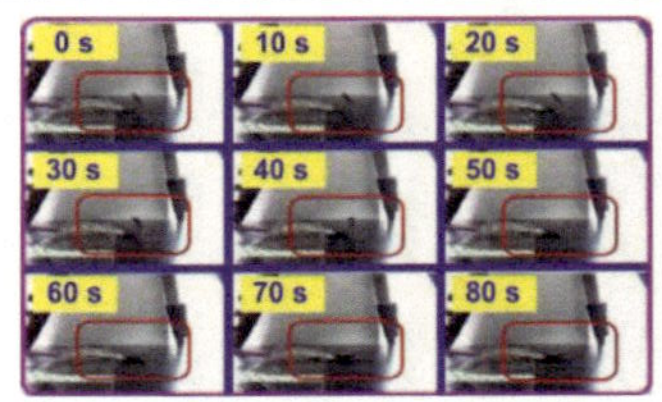 Shape recovery of nanocomposite SMPs induced by Joule heating after been incorporated with 0.03 g of GO and 0.14 g of carbon fiber. 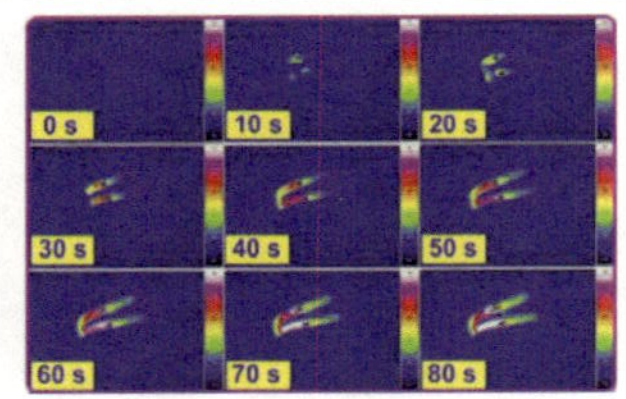 Shape recovery behavior in SMP nanocomposite induced by Joule heating and infrared video camera was utilized for recording.

Author	Substrate	Type of fiber	Recovery ratio (%)	Shape fixity ratio (%)	Comments	Shape memory recovery method/demonstration
Kun Wei [73]	Thermoset Hydro-Epoxy SMP	Glass Fiber, 437H, Taishan Fiberglass	Full recovery	—	• No any noticeable variations in the glass temperature of SMP after the addition of short glass fiber. • Within an amount less than 4.5 wt.% short glass fiber, the hydro-epoxy resin SMP acquired an excellent shape recovery with a 100% recovery after several minutes of different temperature exposures.	 Shape memory properties: A visual perform of SMP after 4.5 wt.% addition of short glass fiber

Table 2.
Summary of various types of fibers and their shape memory effect.

memory characteristics and features are quite limited to due to their abilities of recovering the original shape using the heating source only. Therefore, the reinforcements with micro/nano-fillers and particles or fibers are essentially important to be considered to meet the needed functions and performances. Furthermore, the reinforced SMPs and SMPCs are not only demonstrated a significant mechanical and shape memory properties, but also obtained noble features after being exposure to any electro-or-thermal heating source. Based on the previous researchers, the foreseeable future concepts of SMPs and SMPCs may well-rely on how to adopt the benefits of this kind of properties, in addition to, exceptional attributes as advanced alternatives. On the positive front, the practical applications probability of SMPs and SMPCs are found extensively when displayed in the remarkably distinct application principles which may have seemed in the recent peer-reviewed publications and also patents. Moreover, we believe that the investigation of the advanced features of these types of materials is still in development of new design and/or incorporations. The actual vital, hence, can be found in the finding of substantial beneficial functions wherefore SMPs and SMPCs are enablers or even no less than tremendously excellent substitutes.

Acknowledgements

The author(s) would like to thank the Management and Science University (MSU) for providing the research support under the Seed Research Grant No. SG-451-0518-ISE.

Conflict of interest

The authors declared without any conflict of interest.

Author details

Safaa N. Saud Al-Humairi[1]*, Hasan Sh. Majdi[2], Amir N. Saud Al-Humairi[3] and Mohammed Al-Maamori[4]

1 Faculty of Information Sciences and Engineering, Management and Science University, Shah Alam, Selangor, Malaysia

2 Department of Chemical Engineering, Al-Mustaqbal University Collage, Babylon, Iraq

3 Department of Biomedical Engineering, Al-Mustaqbal University Collage, Babylon, Iraq

4 College of Materials Engineering, University of Babylon, Hilla, Babylon, Iraq

*Address all correspondence to: safaaengineer@gmail.com

References

[1] Liu C, Qin H, Mather P. Review of progress in shape-memory polymers. Journal of Materials Chemistry. 2007;**17**: 1543-1558

[2] Mather PT, Luo X, Rousseau IA. Shape memory polymer research. Annual Review of Materials Research. 2009;**39**:445-471

[3] Lendlein A, Behl M, Hiebl B, Wischke C. Shape-memory polymers as a technology platform for biomedical applications. Expert Review of Medical Devices. 2010;**7**:357-379

[4] Hu J, Zhu Y, Huang H, Lu J. Recent advances in shape–memory polymers: Structure, mechanism, functionality, modeling and applications. Progress in Polymer Science. 2012;**37**:1720-1763

[5] Lagoudas DC. Shape Memory Alloys: Modeling and Engineering Applications. US: Springer; 2008

[6] Hu J. Shape Memory Polymers and Textiles. United Kingdom: Taylor & Francis; 2007

[7] Rousseau IA. Challenges of shape memory polymers: A review of the progress toward overcoming SMP's limitations. Polymer Engineering and Science. 2008;**48**:2075-2089

[8] Leng J, Lan X, Liu Y, Du S. Shape-memory polymers and their composites: Stimulus methods and applications. Progress in Materials Science. 2011;**56**: 1077-1135

[9] Tobushi H, Hara H, Yamada E, Hayashi S. Thermomechanical properties in a thin film of shape memory polymer of polyurethane series. Smart Materials and Structures. 1996;**5**:483

[10] Srivastava S, Biswas A, Senapati S, Ray B, Rana D, Aswal VK, et al. Novel shape memory behaviour in IPDI based polyurethanes: Influence of nanoparticle. Polymer. 2017;**110**:95-104

[11] Li F, Chen Y, Zhu W, Zhang X, Xu M. Shape memory effect of polyethylene/nylon 6 graft copolymers. Polymer. 1998;**39**:6929-6934

[12] Lendlein A, Kelch S. Shape-memory polymers. Angewandte Chemie, International Edition. 2002;**41**: 2034-2057

[13] Kim BK, Lee SY, Xu M. Polyurethanes having shape memory effects. Polymer. 1996;**37**:5781-5793

[14] Lendlein A, Jiang H, Jünger O, Langer R. Light-induced shape-memory polymers. Nature. 2005;**434**:879

[15] Lee BS, Chun BC, Chung Y-C, Sul KI, Cho JW. Structure and thermomechanical properties of polyurethane block copolymers with shape memory effect. Macromolecules. 2001;**34**:6431-6437

[16] Ratna D, Karger-Kocsis J. Recent advances in shape memory polymers and composites: A review. Journal of Materials Science. 2008;**43**:254-269

[17] Wu X, Huang W, Tan H. Characterization of shape recovery via creeping and shape memory effect in ether-vinyl acetate copolymer (EVA). Journal of Polymer Research. 2013; **20**:150

[18] Diani J, Gilormini P, Frédy C, Rousseau I. Predicting thermal shape memory of crosslinked polymer networks from linear viscoelasticity. International Journal of Solids and Structures. 2012;**49**:793-799

[19] Tey S, Huang W, Sokolowski W. Influence of long-term storage in cold

hibernation on strain recovery and recovery stress of polyurethane shape memory polymer foam. Smart Materials and Structures. 2001;**10**:321

[20] Li F, Larock RC. New soybean oil–styrene–divinylbenzene thermosetting copolymers. v. Shape memory effect. Journal of Applied Polymer Science. 2002;**84**:1533-1543

[21] Luo X, Zhang X, Wang M, Ma D, Xu M, Li F. Thermally stimulated shape-memory behavior of ethylene oxide-ethylene terephthalate segmented copolymer. Journal of Applied Polymer Science. 1997;**64**:2433-2440

[22] Liu C, Chun SB, Mather PT, Zheng L, Haley EH, Coughlin EB. Chemically cross-linked polycyclooctene: Synthesis, characterization, and shape memory behavior. Macromolecules. 2002;**35**: 9868-9874

[23] Takahashi T, Hayashi N, Hayashi S. Structure and properties of shape-memory polyurethane block copolymers. Journal of Applied Polymer Science. 1996;**60**:1061-1069

[24] Lin J, Chen L. Study on shape-memory behavior of polyether-based polyurethanes. I. Influence of the hard-segment content. Journal of Applied Polymer Science. 1998;**69**:1563-1574

[25] Gall K, Dunn ML, Liu Y, Finch D, Lake M, Munshi NA. Shape memory polymer nanocomposites. Acta Materialia. 2002;**50**:5115-5126

[26] Wang Z, Liu J, Guo J, Sun X, Xu L. The study of thermal, mechanical and shape memory properties of chopped carbon fiber-reinforced tpi shape memory polymer composites. Polymers. 2017;**9**:594

[27] Pieczyska E, Maj M, Kowalczyk-Gajewska K, Staszczak M, Gradys A, Majewski M, et al. Thermomechanical properties of polyurethane shape memory polymer–experiment and modelling. Smart Materials and Structures. 2015;**24**:045043

[28] Yakacki CM, Satarkar NS, Gall K, Likos R, Hilt JZ. Shape-memory polymer networks with Fe3O4 nanoparticles for remote activation. Journal of Applied Polymer Science. 2009;**112**:3166-3176

[29] Zheng X, Zhou S, Xiao Y, Yu X, Li X, Wu P. Shape memory effect of poly (d, l-lactide)/Fe3O4 nanocomposites by inductive heating of magnetite particles. Colloids and Surfaces. B, Biointerfaces. 2009;**71**:67-72

[30] Meng H, Li G. A review of stimuli-responsive shape memory polymer composites. Polymer. 2013;**54**:2199-2221

[31] Meng Q, Hu J. A review of shape memory polymer composites and blends. Composites Part A: Applied Science and Manufacturing. 2009;**40**: 1661-1672

[32] Lan X, Liu Y, Lv H, Wang X, Leng J, Du S. Fiber reinforced shape-memory polymer composite and its application in a deployable hinge. Smart Materials and Structures. 2009;**18**:024002

[33] Cogswell FN. Thermoplastic Aromatic Polymer Composites: A Study of the Structure, Processing and Properties of Carbon Fibre Reinforced Polyetheretherketone and Related Materials. Oxford, UK: Elsevier; 2013

[34] Seoul C, Kim YT, Baek CK. Electrospinning of poly (vinylidene fluoride)/dimethylformamide solutions with carbon nanotubes. Journal of Polymer Science Part B: Polymer Physics. 2003;**41**:1572-1577

[35] Liu T, Phang IY, Shen L, Chow SY, Zhang W-D. Morphology and mechanical properties of multiwalled carbon nanotubes reinforced nylon-6

composites. Macromolecules. 2004;**37**: 7214-7222

[36] Yu K, Zhang Z, Liu Y, Leng J. Carbon nanotube chains in a shape memory polymer/carbon black composite: To significantly reduce the electrical resistivity. Applied Physics Letters. 2011;**98**:074102

[37] Shaffer MS, Windle AH. Fabrication and characterization of carbon nanotube/poly (vinyl alcohol) composites. Advanced Materials. 1999; **11**:937-941

[38] Raja M, Ryu SH, Shanmugharaj A. Influence of surface modified multiwalled carbon nanotubes on the mechanical and electroactive shape memory properties of polyurethane (PU)/poly (vinylidene diflouride) (PVDF) composites. Colloids and Surfaces A: Physicochemical and Engineering Aspects. 2014;**450**: 59-66

[39] Abishera R, Velmurugan R, Gopal KN. Reversible plasticity shape memory effect in epoxy/CNT nanocomposites-a theoretical study. Composites Science and Technology. 2017;**141**:145-153

[40] Qi X, Xiu H, Wei Y, Zhou Y, Guo Y, Huang R, et al. Enhanced shape memory property of polylactide/thermoplastic poly (ether) urethane composites via carbon black self-networking induced co-continuous structure. Composites Science and Technology. 2017;**139**:8-16

[41] Lu H, Yao Y, Huang WM, Leng J, Hui D. Significantly improving infrared light-induced shape recovery behavior of shape memory polymeric nanocomposite via a synergistic effect of carbon nanotube and boron nitride. Composites Part B: Engineering. 2014; **62**:256-261

[42] Zhang H, Zhao Y. Polymers with dual light-triggered functions of shape memory and healing using gold nanoparticles. ACS Applied Materials & Interfaces. 2013;**5**:13069-13075

[43] Pilate F, Toncheva A, Dubois P, Raquez J-M. Shape-memory polymers for multiple applications in the materials world. European Polymer Journal. 2016; **80**:268-294

[44] Wang L, Wang W, Di S, Yang X, Chen H, Gong T, et al. Silver-coordination polymer network combining antibacterial action and shape memory capabilities. RSC Advances. 2014;**4**:32276-32282

[45] Lu H, Liang F, Gou J, Leng J, Du S. Synergistic effect of Ag nanoparticle-decorated graphene oxide and carbon fiber on electrical actuation of polymeric shape memory nanocomposites. Smart Materials and Structures. 2014;**23**:085034

[46] Luo H, Li Z, Yi G, Zu X, Wang H, Wang Y, et al. Electro-responsive silver nanowire-shape memory polymer composites. Materials Letters. 2014;**134**: 172-175

[47] Gojny FH, Wichmann MH, Fiedler B, Kinloch IA, Bauhofer W, Windle AH, et al. Evaluation and identification of electrical and thermal conduction mechanisms in carbon nanotube/epoxy composites. Polymer. 2006;**47**: 2036-2045

[48] Hribar KC, Metter RB, Burdick JA. Novel nano-composite biomaterials that respond to light. In: Engineering in Medicine and Biology Society, 2009 EMBC 2009 Annual International Conference of the IEEE. Piscataway, US: IEEE; 2009. pp. 2409-2411

[49] Zheng Y, Li J, Lee E, Yang S. Light-induced shape recovery of deformed shape memory polymer micropillar arrays with gold nanorods. RSC Advances. 2015;**5**:30495-30499

[50] Zhang H, Xia H, Zhao Y. Light-controlled complex deformation and motion of shape-memory polymers using a temperature gradient. ACS Macro Letters. 2014;**3**:940-943

[51] Zhang H, Zhang J, Tong X, Ma D, Zhao Y. Light polarization-controlled shape-memory polymer/gold nanorod composite. Macromolecular Rapid Communications. 2013;**34**:1575-1579

[52] Liang C, Rogers C, Malafeew E. Investigation of shape memory polymers and their hybrid composites. Journal of Intelligent Material Systems and Structures. 1997;**8**:380-386

[53] Rong MZ, Zhang MQ, Liu Y, Yang GC, Zeng HM. The effect of fiber treatment on the mechanical properties of unidirectional sisal-reinforced epoxy composites. Composites Science and Technology. 2001;**61**:1437-1447

[54] Botelho EC, Silva RA, Pardini LC, Rezende MC. A review on the development and properties of continuous fiber/epoxy/aluminum hybrid composites for aircraft structures. Materials Research. 2006;**9**: 247-256

[55] Ohki T, Ni Q-Q, Iwamoto M. Creep and cyclic mechanical properties of composites based on shape memory polymer. Science and Engineering of Composite Materials. 2004;**11**:137-148

[56] Pretsch T. Review on the functional determinants and durability of shape memory polymers. Polymers. 2010;**2**: 120-158

[57] Zhang C-S, Ni Q-Q. Bending behavior of shape memory polymer based laminates. Composite Structures. 2007;**78**:153-161

[58] Francis W, Lake M, Mallick K, Freebury G, Maji A. Development and testing of a hinge/actuator using elastic memory composites. In: 44th AIAA/ASME/ASCE/AHS/ASC Structures, Structural Dynamics, and Materials Conference. 2003. p. 1496

[59] Lake MS, Campbell D. The fundamentals of designing deployable structures with elastic memory composites. In: Aerospace Conference, 2004 Proceedings 2004 IEEE. Vol. 2004. IEEE. pp. 2745-2756

[60] Gall K, Mikulas M, Munshi NA, Beavers F, Tupper M. Carbon fiber reinforced shape memory polymer composites. Journal of Intelligent Material Systems and Structures. 2000; **11**:877-886

[61] Leng J, Du S. Shape-Memory Polymers and Multifunctional Composites. United States: CRC Press; 2010

[62] Abrahamson ER, Lake MS, Munshi NA, Gall K. Shape memory mechanics of an elastic memory composite resin. Journal of Intelligent Material Systems and Structures. 2003;**14**:623-632

[63] Li F, Scarpa F, Lan X, Liu L, Liu Y, Leng J. Bending shape recovery of unidirectional carbon fiber reinforced epoxy-based shape memory polymer composites. Composites Part A: Applied Science and Manufacturing. 2019;**116**: 169-179

[64] Iregui A, Irusta L, Llorente O, Martin L, Calvo-Correas T, Eceiza A, et al. Electrospinning of cationically polymerized epoxy/polycaprolactone blends to obtain shape memory fibers (SMF). European Polymer Journal. 2017;**94**:376-383

[65] Robinson P, Bismarck A, Zhang B, Maples HA. Deployable, shape memory carbon fibre composites without shape memory constituents. Composites Science and Technology. 2017;**145**:96-104

[66] Guo J, Wang Z, Tong L, Lv H, Liang W. Shape memory and

thermo-mechanical properties of shape memory polymer/carbon fiber composites. Composites Part A: Applied Science and Manufacturing. 2015;**76**: 162-171

[67] Li G, Ajisafe O, Meng H. Effect of strain hardening of shape memory polymer fibers on healing efficiency of thermosetting polymer composites. Polymer. 2013;**54**:920-928

[68] Li G, Zhang P. A self-healing particulate composite reinforced with strain hardened short shape memory polymer fibers. Polymer. 2013;**54**: 5075-5086

[69] Herath H, Epaarachchi J, Islam M, Al-Azzawi W, Leng J, Zhang F. Structural performance and photothermal recovery of carbon fibre reinforced shape memory polymer. Composites Science and Technology. 2018;**167**:206-214

[70] Tan Q, Liu L, Liu Y, Leng J. Thermal mechanical constitutive model of fiber reinforced shape memory polymer composite: Based on bridging model. Composites Part A: Applied Science and Manufacturing. 2014;**64**: 132-138

[71] Hallad SA, Banapurmath NR, Hunashyal AM, Shettar AS, Ayachit NH, Mruthunjaya A, et al. Experimental investigation for graphene and carbon fibre in polymer-based matrix for structural applications. Journal of Applied Research and Technology. 2017; **15**:297-302

[72] Lu H, Liu Y, Gou J, Leng J, Du S. Synergistic effect of carbon nanofiber and carbon nanopaper on shape memory polymer composite. Applied Physics Letters. 2010;**96**:084102

[73] Wei K, Zhu G, Tang Y, Li X, Liu T, Niu L. An investigation on shape memory behaviours of hydro-epoxy/ glass fibre composites. Composites Part B: Engineering. 2013;**51**:169-174

Chapter 4

Magnetorheological Elastomers: Materials and Applications

Taixiang Liu and Yangguang Xu

Abstract

Magnetorheological elastomers (MREs) are a type of soft magneto-active rubber-like material, whose physical or mechanical properties can be altered upon the application of a magnetic field. In general, MREs can be prepared by mixing micron-sized magnetic particles into nonmagnetic rubber-like matrices. In this chapter, the materials, the preparing methods, the analytical models, and the applications of MREs are reviewed. First, different kinds of magnetic particles and rubber-like matrices used to prepare MREs, as well as the preparing methods, will be introduced. Second, some examples of the microstructures, as well as the microstructure-based analytical models, of MREs will be shown. Moreover, the magnetic field-induced changes of the macroscopic physical or mechanical properties of MREs will be experimentally given. Third, the applications of MREs in engineering fields will be introduced and the promising applications of MREs will be forecasted. This chapter aims to bring the reader a first-meeting introduction for quickly knowing about MREs, instead of a very deep understanding of MREs.

Keywords: magnetorheological elastomer, composite material, microstructure, vibration reduction, energy absorption

1. Introduction

Magnetorheological elastomers (MREs) are a type of soft particle-reinforced magneto-active rubber-like composite material, whose physical or mechanical properties can be altered upon the application of a magnetic field [1–5]. MREs can be usually prepared by mixing micron-sized magnetic particles into nonmagnetic rubber-like matrices. In the presence of a magnetic field, MREs exhibit a magnetorheological effect providing a field-dependent physical or mechanical property, for example, a controllable modulus, due to the sensitive response of the magnetic particles to the field. While the field is removed, MREs will reclaim their original, natural property. It is believed that the embryo of MREs is firstly reported by Rigbi and Jilken [1, 6] in 1983, although the discovery of the basic magnetorheological effect can be historically retrospected to the 1940s for magnetic fluid [7]. MREs can be regarded as a solid-state analog to magnetorheological fluids (MRFs) [8–12]. In general, MREs exhibit a unique field-dependent material property when exposed to a magnetic field, and can overcome major issues faced in MRFs, for example, the deposition of iron particles, sealing problems, and environmental contamination. Such advantages offer MREs great potential for designing intelligent devices to be used in various engineering fields, especially in fields that involve vibration

IntechOpen

reduction, isolation, and absorption [1, 13–16]. Recently, the study of the sensing behavior of MREs explosively emerged, for example, for sensing mechanical and magnetic signals [17–22].

From the first report of MREs with soft ferrite particles filling into natural rubber [6], the research of the materials and related preparing methods of MREs develops very quickly. Briefly speaking, MREs consist of three basic components: magnetic particles, nonmagnetic elastic matrices, and additives. For the magnetic particles, higher permeability, higher saturation magnetization, and lower remanent magnetization are highly desirable for obtaining stronger magnetic field-sensitive effect. At present, the micron-sized carbonyl iron powder invented by BASF in 1925 is widely used [23, 24]. For the elastic matrix, there are lots of polymeric rubbers, for example, natural rubber [25, 26], silicone rubber [27–31], poly dimethylsiloxane (PDMS) [32–34], etc., for consideration. According to need, high modulus or low modulus can be chosen, but that the magnetic particles can be locked in the matrix in the absence or presence of a magnetic field is a basic requirement. For the additives, they are determined according to the choice of the particles and the matrix, and silicone oil is usually used as an additive in the fabrication of MREs [1]. Depending on the choice of the matrix, the preparing methods of MREs are many and various, and a high-temperature or room-temperature vulcanization curing method is usually used. Attributed to applying a magnetic field in the curing process, MREs can be prepared with anisotropic particle-formed microstructure. This kind of MREs is called anisotropic MREs. When no field is applied during the curing process, prepared MREs have isotropic particle-formed microstructure and this kind of MREs is called isotropic MREs. It is worth mentioning that the properties of isotropic MREs can differ much from those of anisotropic MREs.

Due to magnetic field-sensitive response of the magnetic particles, the material properties of MREs can be altered by using a magnetic field. For MREs, the magnetorheological effect is defined as the ratio of the value increment of a property at a measured magnetic field to the initial value of that property at zero magnetic field. In most studies, the distinctive change of the storage or loss modulus of MREs is a common concern. The magnetorheological effect is characterized by the ratio of modulus increment ΔG at a measured magnetic field to the initial modulus G_0, i.e., $\Delta G/G_0$. For example, **Figure 1** gives the magnetic field-dependent shear storage modulus of MRE. The initial modulus G_0 of the MRE is about 0.6 MPa resulting from the initial state of MRE materials. When applying a magnetic field,

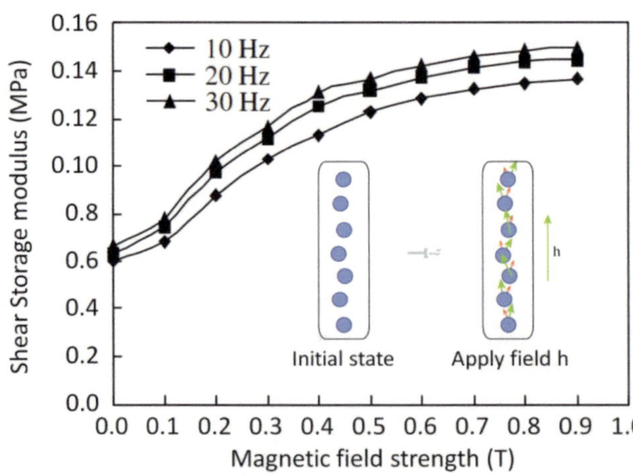

Figure 1.
The magnetic field-dependent shear storage modulus of MREs (modified from Refs. [13, 35]).

the interaction between the magnetic particles of MRE will occur and result in the alteration of the storage modulus. The modulus of this MRE can reach 1.5 MPa in the presence of field and thus the relative change of modulus can be larger than 100%. It should be noticed that, besides the magnetic field, the temperature [36–39], the relative humidity [40], and even the γ radiation [41] can influence the physical or mechanical properties of MREs. Based on the magnetic field-induced change of the physical or mechanical properties of MREs, lots of magnetorheological devices have been designed. In the late 1990s, Ginder et al. [2, 42] and Carlson et al. [3] had done the pioneering works and suggested controllable-stiffness components and electrically-controllable mounts based on MREs. Recently, Li et al. [1] and Ubaidillah et al. [43] have presented state-of-the-art reviews on magnetorheological elastomer devices. From these reviews, one can conclude that MREs can be used in many devices including but not limited to vibration absorbers, vibration isolators, sensors, controllable valves, and adaptive beam structures.

In this chapter, the materials, the preparing methods, the analytical models, and the applications of MREs will be reviewed. In the following section (Section 2), the materials, i.e., the magnetic particles, the nonmagnetic matrix and the additives, as well as the preparing methods of MREs will be introduced. In Section 3, the microstructures of prepared MREs, the analytical models of MREs, and some typical macroscopic properties of MREs will be presented. The relationship between the microstructures and the macroscopic properties will be qualitatively discussed. Then the applications of MREs will be briefly introduced in Section 4. Finally, Section 5 will give a summary of this chapter.

2. Materials

2.1 Magnetic particles

To the magnetic particles, higher permeability, higher saturation magnetization, and lower remanent magnetization are highly desirable for obtaining stronger magnetic field-sensitive effect. Among a variety of magnetic particle materials, micrometer-sized carbonyl iron (CI) powder is currently widely used as a magnetic particle for preparing MREs. For a quick recognition, **Figure 2(a)** shows an example of the macroscopic image of CI powder. It shows that the CI powder is a very fine powder material. **Figure 2(b)** and **(c)** shows the scanning electron microscopy (SEM) images of CI powder (Type CN, produced by BASF SE Inc.) with different magnifications. The diameter of this kind of CI powder is several micrometers.

In general, the size of the magnetic particles can range from several micrometers to hundreds of micrometers [47, 48]. **Figure 3** gives the size distribution of CI powder from experimental test. The size distribution can be analytically modeled by a lognormal distribution model as the following equation (Eq. (1)) shows.

$$P(d) = \frac{1}{d\sigma\sqrt{2\pi}} \exp\left[-\frac{(\ln(d) - \mu)^2}{2\sigma^2}\right], \tag{1}$$

in which $P(d)$ is the probability density distribution function of the diameter of CI powder material. d is the diameter of the CI particle. μ and σ are the expectation and variance of $\ln(d)$. The tap density of CI powder is usually about 3.0 g/cm^3 and the real density of CI powder is about 7.0 g/cm^3.

Beside the size distribution of CI powder, the magnetic property of CI powder draws more attention of researchers. Higher permeability, higher saturation magnetization, and less remanent magnetization of magnetic particles are always highly desirable for

Figure 2.
Images of carbonyl iron powder. The upper subfigure (a) shows the macroscopic image [44] in daily view and the lower two subfigures (b and c) show the SEM images with different magnifications [45].

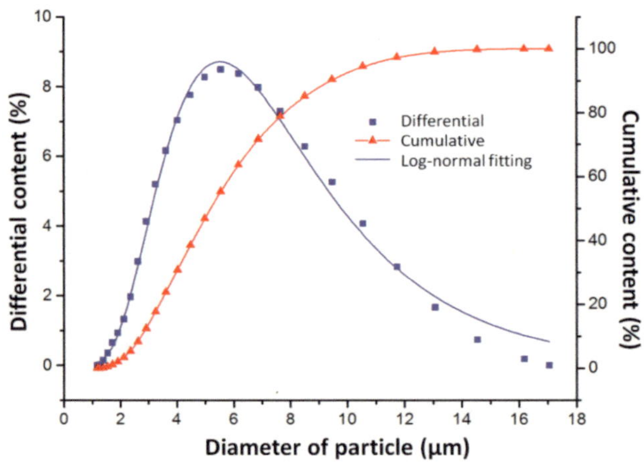

Figure 3.
The size distribution of carbonyl iron powder (Type CIP-CN) [46].

obtaining stronger magnetic field-sensitive effect. As is shown in **Figure 4**, CI powder shows very high permeability, saturation magnetization, and very little remanent magnetization. The value of saturation magnetization can reach more than 600 kA/m and there is little remanent magnetization when magnetic field is removed. This mainly results from the fact that the content of Fe element in CI powder is usually more than 97.5% in weight fraction. Attributing to the excellent magnetic property, CI powder is widely used for fabricating materials including but not limited to MREs.

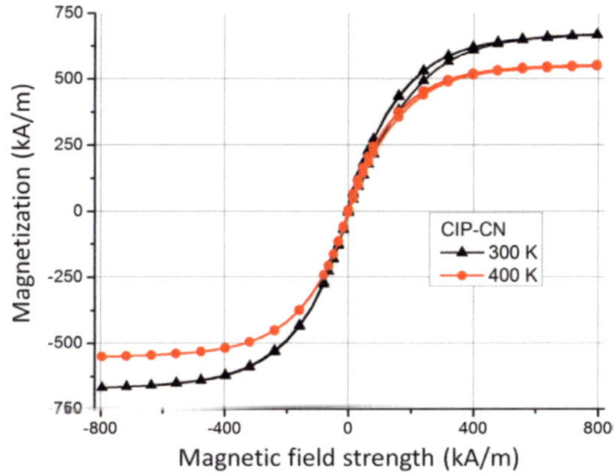

Figure 4.
The magnetic hysteresis loop of carbonyl iron powder (Type CIP-CN) at different temperatures [49].

2.2 Elastic matrices and additives

A basic requirement of elastic matrices for fabricating MREs is that the matrices have soft elastic property, meaning that the matrices can stably hold the magnetic particles under no magnetic field and have a finite deformation under a magnetic field. For the elastic matrices, there are lots of polymeric rubbers that can be considered as candidates, for example, silicone rubber [50], natural rubber [51], butadiene rubber [52], butyl rubber [53], polyurethane [54], polydimethylsiloxane [55], epoxy [56], etc. For instantly having a basic recognition of the matrices, the following **Figure 5** gives some examples of image and applications of widely used silicone rubber.

The modulus of these matrices differs much from each other. For example, under normal conditions, the modulus of silicone rubber can be lower than 1.0 MPa [27]. That of natural rubber often reaches several MPa [25]. The shear modulus of PDMS varies with preparation conditions, but is typically in the range of 0.1–3.0 MPa [55]. The modulus of polyurethane can range from 0.01 MPa to several hundred MPa, attributed to its raw materials from fluid-like to solid-like [54]. Among a large amount of rubbers, silicone rubber compounds have characteristics of both inorganic and organic materials, and offer a number of advantages not found in other organic rubbers. From **Figure 5**, one can know that silicone rubbers have mechanically low modulus and good chemical stability and are nontoxic, nonpolluting, and human-body-friendly in daily use. As is shown in **Figure 6**, compared to the modulus of other rubbers, the modulus of silicone rubber is much lower within a large range of temperature. Besides, the thermal conductivity of silicone rubber can vary in a wide range. Based on the above-mentioned properties, silicone rubber can be chosen as an ideal soft elastic matrix for preparing MREs and is widely used in fabricating MREs. The other rubber matrices, with some unique mechanical or physical properties for special usage, can also been used in fabricating MREs according to need.

Besides the magnetic particles and the elastic matrices, additives are also key components for preparing MREs. Silicone oil is usually used as an additive in material fabrication of MREs. When the molecules of the silicone oil enter the matrix, the gaps between the matrix molecules are increased, and the conglutination of the molecules is decreased. Apart from increasing the plasticity and fluidity of the matrix, the additives can average the distribution of the internal stress in the materials, which makes a stable material property for MR elastomer materials [1, 58]. The other additives include but are not limited to carbon black [59–61], carbon nanotubes [62–65], silver nanowire [20], Rochelle salt [30], gamma-ferrite additives [66], etc.

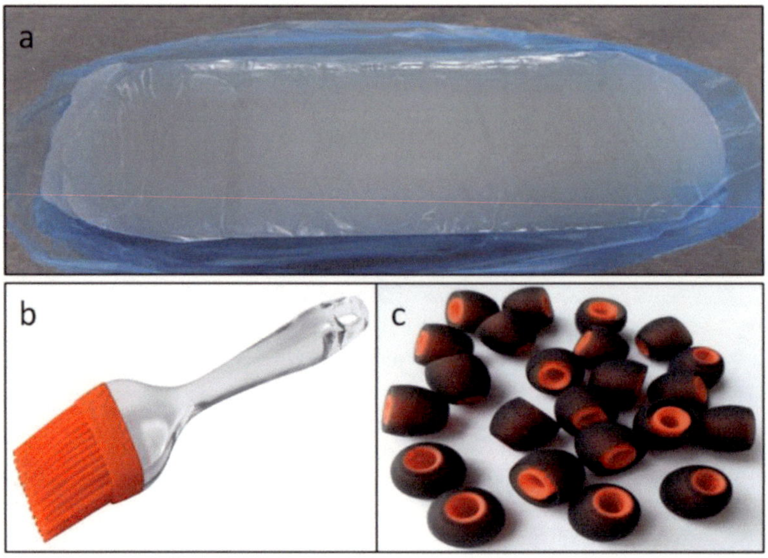

Figure 5.
The images of bulk material and related daily applications of silicone rubber [48]. The upper subfigure (a) shows an image of bulk material of silicone rubber. The lower subfigures (b and c) give the examples of brush and earplug made from soft silicone rubber.

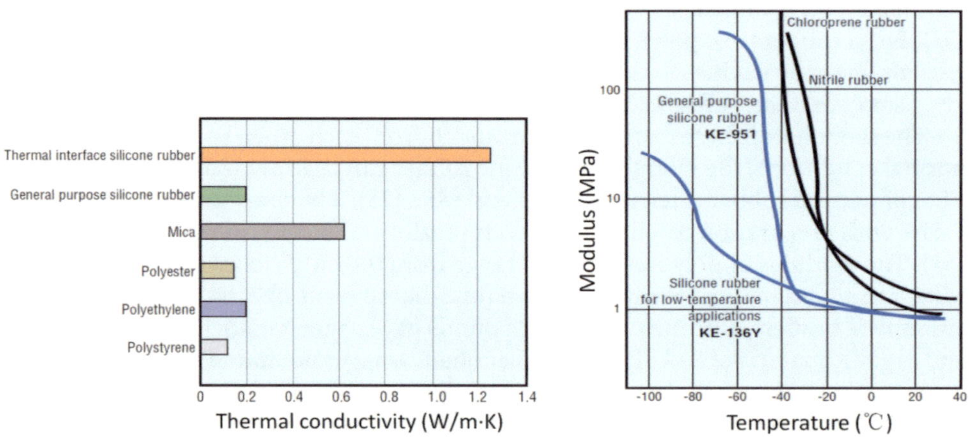

Figure 6.
The thermal conductivity and the modulus of silicone rubber (produced by Shin-Etsu Chemical Co., Ltd. Japan) [57]. The thermal conductivity of silicone rubber can vary in a wide range. Comparing to the modulus of other rubbers, the modulus of silicone rubber is much lower.

2.3 Preparing methods

A simplified illustration of the processing for preparing MREs is shown in **Figure 7**. Usually, the magnetic particles and the matrix are mechanically mixed with some additives into a mixture. The mixture has a very low yield stress, meaning that the mixture can easily deform and usually creep with itself. Then the mixture vulcanizes at room temperature (called room-temperature vulcanizing, RTV [4]) or high temperature (called high-temperature vulcanizing, HTV [45]) higher than 120°C. During the vulcanizing, in case of applying a magnetic field, the magnetic particles can move in the matrix and gradually aggregate forming chain-like structures along the direction of the field. After the magnetic field-assisted curing,

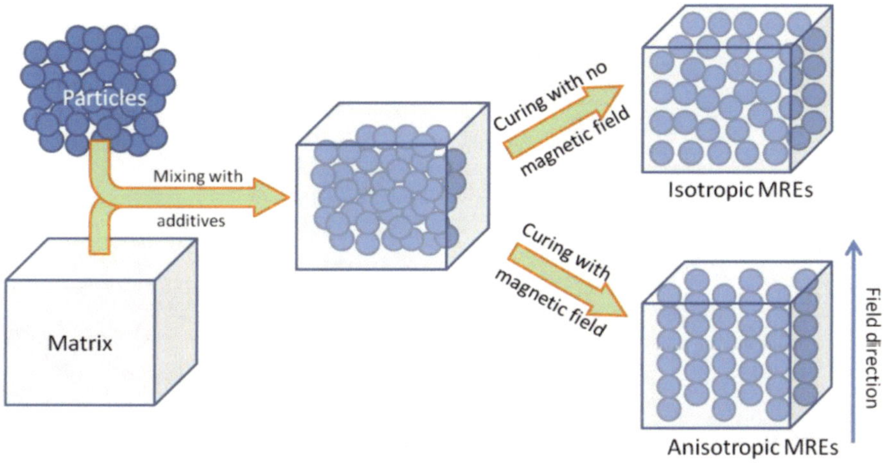

Figure 7.
Illustration of the processing for preparing MREs. The magnetic particles and the matrix are mixed with additives into a mixture. When the mixture is cured with no external magnetic field, the mixture will be cured into isotropic MREs. However, in the case of the mixture curing under a uniform magnetic field, the mixture will be cured into anisotropic MREs.

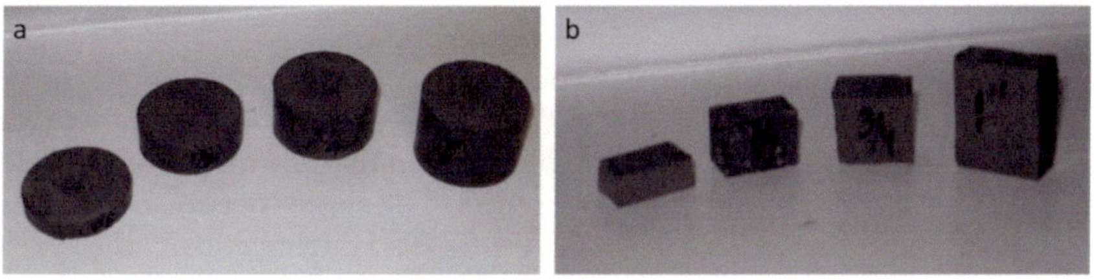

Figure 8.
Prepared cylindrical (a) and block (b) MREs with different thicknesses from 6.35 to 2.54 cm [67].

anisotropic MREs, meaning that the magnetic particles form chain-like microstructures in MREs, can be prepared. When curing with no magnetic field, the magnetic particles will disperse uniformly in the matrix after vulcanization and thus isotropic MREs are prepared. **Figure 8** shows some images of prepared MRE samples.

3. Microstructures and macroscopic properties of MREs

3.1 Microstructures of MREs

As a basic concern in the research of MREs, there are lots of studies focusing on the microstructures of MREs. Almost every report about a newly prepared MRE will show its microstructure. **Figure 9** gives some typical SEM images with different times of magnification of the microstructures of carbonyl iron powder-embedded, natural rubber-based MREs samples. The volume fractions of iron particles for all samples are 11%. **Figure 9(a)** shows the images of a MRE sample cured with no magnetic field. It shows that the carbonyl iron particles randomly and uniformly disperse in the matrix. These two images are typical images showing the microstructures of isotropic MREs. The other images, i.e. **Figure 9(b–f)**, show the microstructures of anisotropic MREs cured with magnetic field. As is shown, the magnetic particles will aggregate forming chain-like microstructures. The stronger the magnetic field intensity is when curing, the longer and thicker the magnetic particle-formed chains, as

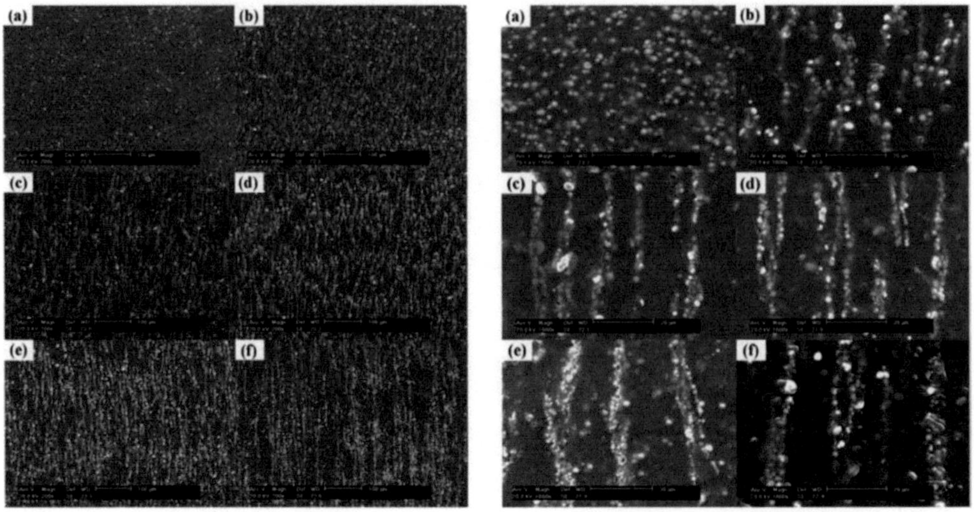

Figure 9.
SEM images with 200 times (left series of subfigures) and 1600 times (right series of subfigures) magnification of MREs prepared under a magnetic flux intensity B of (a) 0 mT, (b) 200 mT, (c) 400 mT, (d) 600 mT, (e) 800 mT, and (f) 1000 mT [68].

the magnetic interaction of neighbor particles is stronger. When the magnetic field is not strong enough, for example, 200 mT, the magnetic particles in the matrix can only move within a small width of range, resulting in that the magnetic particles can only form some short chain-like microstructures. The spaces between these chains are small. With the enhancement of the magnetic field during curing, the spaces will get wider and the anisotropy of MREs will get higher, implying that the properties of MREs will get more anisotropic.

In recent years, the tendency to use quantitative methods instead of a qualitative analysis for structural investigation of the structure of nano- and microstructures has been increasing. Tomographic data open the possibility to achieve detailed quantitative data using techniques of digital image processing [69, 70]. To experimentally study the three-dimensional (3D) microstructures of MREs, Borin's group [71, 72] firstly used the X-ray micro-computed tomography (XμCT) method to investigate the microstructures of MREs.

Balasoiu et al. [70] allowed a detailed structural analysis of both isotropic MREs and anisotropic MREs. **Figure 10** shows the exemplary XμCT images of silicone rubber-based isotropic and anisotropic MRE samples. The iron particles in these MREs have an average particle size of approximately 35 μm. With XμCT images, single microparticles and aggregates as well as their spatial position can be identified. As can be seen, the particles are distributed homogeneously in the isotropic MRE sample and form chain-like structures in the anisotropic one. **Figure 11** shows two extracted particle-formed columns from the reconstructed XμCT image. In this figure, the direction of the magnetic field was parallel to the longitudinal axis of the cylindrical shape holders and to the gravitational force. From this XμCT image, one can see how the particle chain forms and can go further to model MREs based on the image. Recently, the motion of particles in MREs was investigated by using XμCT [73]. It has been shown that XμCT is a powerful technique to investigate the inner structure of macroscopic samples without destroying the specimen. The XμCT technique also achieves high spatial resolution and allows the derivation of valuable local and statistical information, such as particle size and position, from the reconstructed 3D images. Furthermore, the nondestructive XμCT investigations

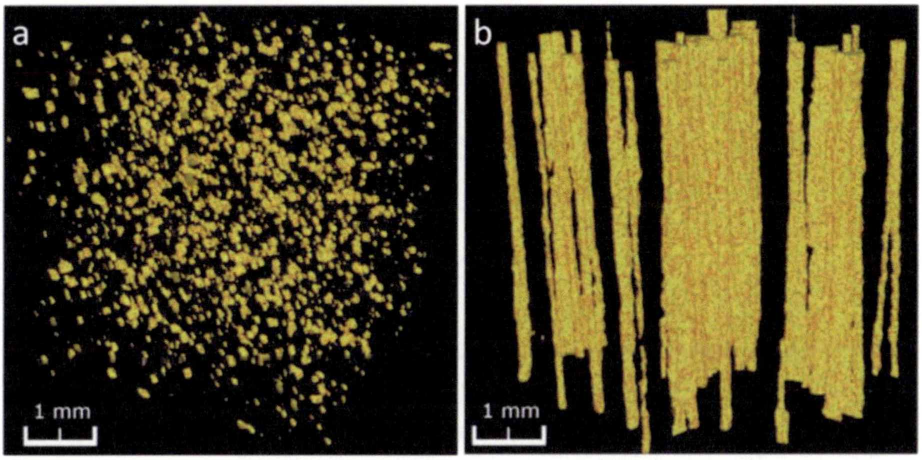

Figure 10.
Exemplary XμCT images of the isotropic (a) and anisotropic (b) MRE samples with ~5 wt% magnetic particles. [67].

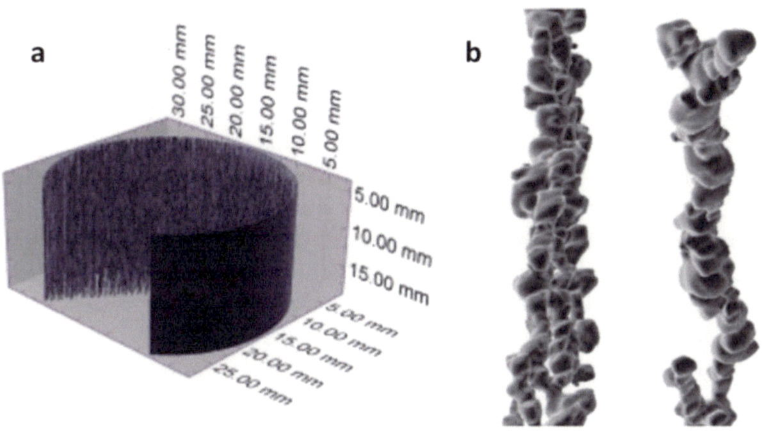

Figure 11.
Magnetorheological elastomer: (a) part of the reconstructed image of a MRE sample and (b) extracted columns from the reconstructed tomography image. The directions of magnetic field and gravity are downward.

of the 3D microstructures of MREs introduce the possibility of investigating the influence of the number and size of columns on the macroscopic mechanical and rheological properties of MREs.

3.2 Analytical models

The study of the microstructure-based analytical model of MREs is always a key work for deeply knowing about MREs. Some pioneering works had been done studying the model of MREs in the late 1990s [74, 75] and a single-chain model was proposed assuming that the MREs are fully filled with single chains (**Figure 12(a)**). From these studies, it was reported that the optimum particle volume fraction for the largest fractional change in modulus at saturation is predicted to be 27%. Calculations of the zero-field shear modulus perpendicular to the chain axis indicate that it does not exceed the modulus of a filled elastomer with randomly dispersed particles of the same concentration. In 2010, Li and Zhang proposed bimodal particle-based chain-model of MREs (**Figure 12(b)**) [76]. In their work, theoretical and experimental

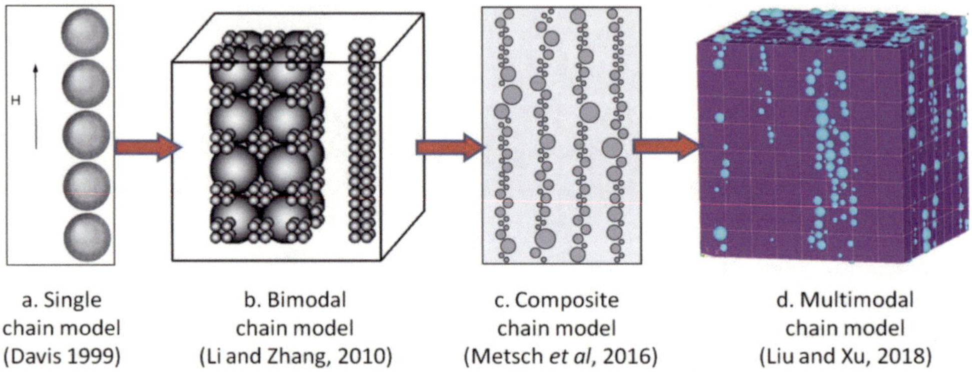

Figure 12.
The evolution of the model of MREs from (a) single chain model [73], to (b) bimodal chain model [74], to (c) composite chain model [75], and to the latest (d) multimodal chain model [77].

studies of the mechanical performance and magnetorheological effects of MREs fabricated with mixtures of large and small particles were performed and an effective permeability model was developed to theoretically analyze the MR effect of bimodal particle-based MR elastomers. Six years later, a composite chain model (**Figure 12(c)**) was proposed [77, 78]. In this work, a particle chain composed of multiple kinds of particles with different sizes of diameter was introduced and a modeling strategy which accounts for elastic constituents and a nonlinear magnetization behavior of the particles is pursued. Most recently, a 3D multimodal chain model [79] has been proposed. In this model, magnetic particles with log-normal size distribution of diameter were introduced as fillers in soft elastic matrix. At the same time, a finite element model was built according to this model (**Figure 12(d)**). With the finite element model, one can computationally study the macroscopic physical or mechanical properties of MREs. In addition, as a basic issue, the study of the interaction between two magnetic particles still keeps developing [80–82]. Moreover, besides the microstructure-based analytical model of MREs, the phenomenological continuous medium-based models were also studied [83–93]. These works focus on theoretically and/or experimentally studying the magneto-viscoelastic models for MREs.

3.3 Macroscopic properties

The most important characterization of MREs is that their macroscopic physical or mechanical properties can be altered upon the application of a magnetic field. For a long time, most studies have focused on the magnetic field-induced changes of the modulus or damping of MREs [1]. The shear storage/loss modulus or damping property of MREs can be measured by dynamic mechanical analyzer (DMA) or rheometer. As examples, **Figure 13** shows the magnetic field-dependent shear storage modulus and damping of MRE samples. For some natural rubber-based MRE samples with different weight fraction of carbonyl iron particles, one can find that the magnetic field-induced change of their shear storage modulus can reach near or above two times of their initial magnitude. Moreover, as a characterizing of the damping property of MREs, the relationship between shear stress and shear strain, under various magnetic field strengths of silicone rubber-based MRE, is shown in **Figure 13** (right). The results show that such MREs have controllable damping properties. The increase of the stress-strain loop area with magnetic field demonstrates that the damping capacity of MREs is a function of applied magnetic field. These field-dependent mechanical properties make MREs much promising in many engineering fields, especially in vibration reduction.

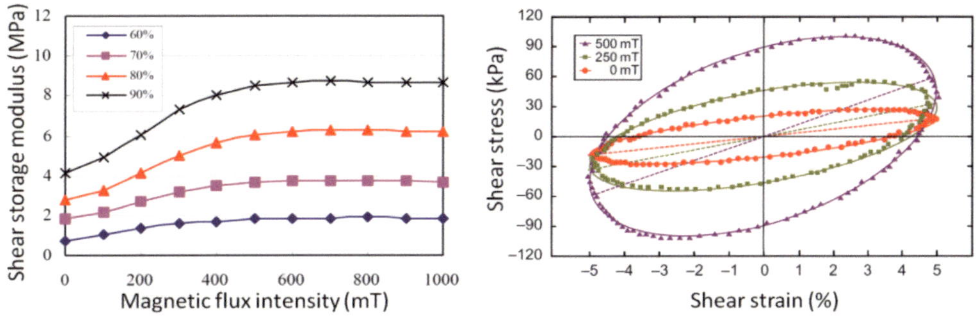

Figure 13.
Left: magnetic field-dependent shear storage modulus of different natural rubber-based MRE samples with 60, 70, 80, and 90% of carbonyl iron powder in weight fraction [25]. Right: stress-strain relationship of a silicone rubber-based MRE sample at various magnetic fields [94].

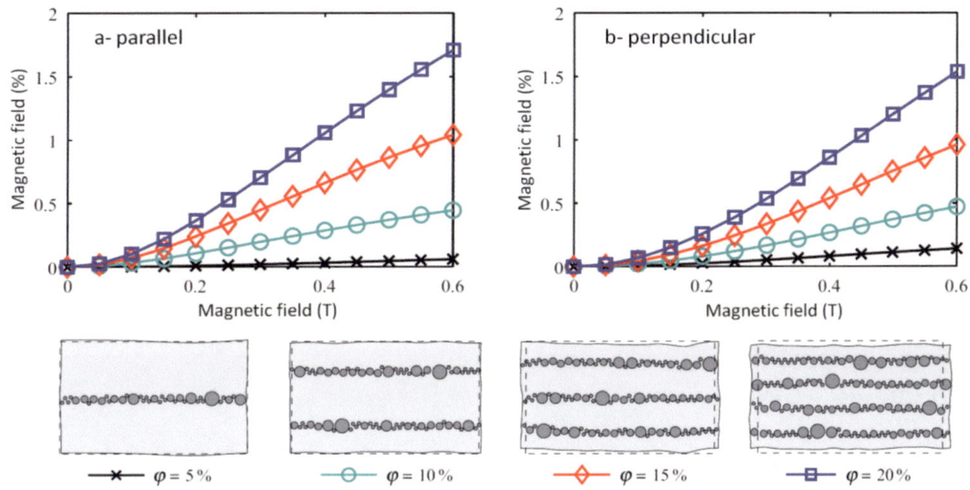

Figure 14.
Simulated magnetostriction in MRE with structured particle distributions: (a) effective magnetostrictive strains in the case of the particle chain being parallel to an applied magnetic field and (b) results for particle chain being perpendicular to an applied magnetic field. [75].

Magnetostriction of MREs is also a key concern when studying MREs [75, 95–97]. **Figure 14** shows the simulated magnetostriction in MRE with different volume fractions of structured particle distributions. It shows that the magnetostriction of MREs is magnetic field-dependent. The intenser the field is and the higher the volume fraction of magnetic particle is, the larger the magnetostriction is. Further, full-field magnetostriction/deformation of a silicone rubber-based MRE under uniform magnetic field has been studied [98]. It shows that both isolated particles and grouped particles result in the concave-convex deformation of the MRE sample. Recently, the magnetic field-dependent electrical conductivity of MREs was emergently studied [99–105]. These field-dependent properties make MREs much promising in actuating and sensing.

4. Applications

In 1993, Kordonsky pointed out that magnetorheological effect can be a base of new devices and technologies [106]. Years later, Carlson and Jolly gave an introduction of magnetorheological devices [3]. By possessing variable physical or mechanical properties when subjected to a magnetic field, MREs are natural candidates to be developed in many applications. In 2014, Li et al. [1] presented a state-of-the-art review on MRE-based devices. From this review, one can find that MREs can be used in many devices including but not limited to vibration absorbers, vibration isolators, sensors, controllable valves, and adaptive beam structures.

4.1 Vibration absorbers

Ginder et al. [107] firstly constructed a simple one-degree-of-freedom mass-spring system—an adaptive tuned vibration absorber—that utilizes MREs as variable-spring-rate elements. After that, the research on vibration MRE-based absorbers developed quickly (e.g., Refs. [13, 108–112]). **Figure 15** shows the sketch of a designed MRE-based vibration absorber composed of a semi-active vibration absorption unit and a passive vibration isolation unit. The vibration absorption unit is composed of a magnetic conductor, a shearing sleeve, a bobbin core, an electromagnetic coil winding, and a circular cylindrical MRE vulcanized between the shearing sleeve and the bobbin core. The magnetic conductor, the bobbin core, and the electromagnetic coil are supported on the shearing sleeve through the MRE. The magnetic conductor and the bobbin core are connected by a bolt, and the shearing sleeve is fixed to the lower housing. The outer surface of the shearing sleeve is in clearance fit with the inner surface of the magnetic conductor, and the magnetic conductor can move vertically along the shearing sleeve. The MRE works in pure shear mode, and the magnetic conductor, the bobbin core, and the electromagnetic coil form the dynamic mass of the MRE-based vibration absorber together. The proposed MRE-based vibration absorber can absorb the vibration energy and thus reduce vibration.

4.2 Vibration isolators

Vibration isolators are devices which can isolate an object, such as a piece of equipment, from the source of vibration. Vibration isolators can be categorized into two groups: base isolation and force isolation, and the isolating modes have active and passive vibration isolation [113]. In Ref. [1], Li et al. had given a review on the application of vibration isolators for mechanical engineering and civil engineering. There are many works that focused on the study of MRE-based vibration isolators, for example, Refs. [16, 91, 92, 114–118]. **Figure 16** shows an example

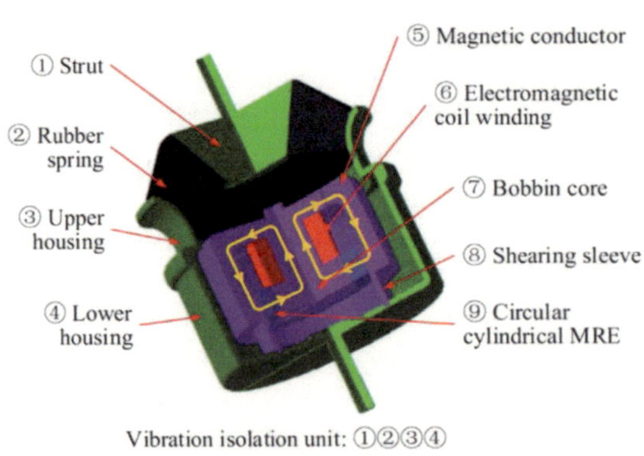

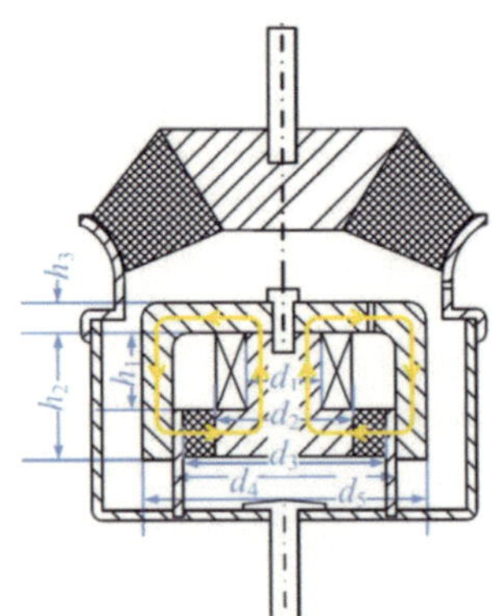

Figure 15.
The 3D drawing (left) and the schematic representation (right) of a MRE-based dynamic vibration absorber [109].

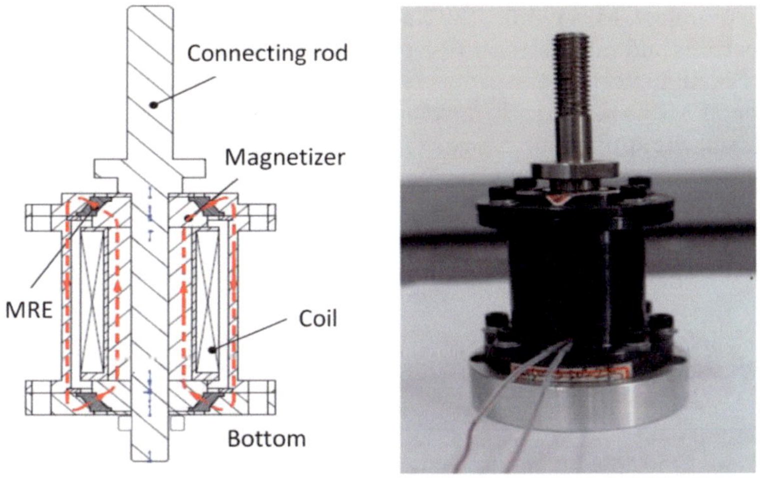

Figure 16.
MRE-based isolator: cross-sectional view of designed layout (left) and fabricated prototype (right) [116].

of the designed layout and prototype of MRE-based isolator working in squeeze/elongation-shear mode. It shows that the initial vertical stiffness and damping coefficient of the magnetorheological elastomer isolator are 1.14×10^6 N/m and 495.8 N·s/m, respectively. The relative increase in stiffness and damping is 66.57% and 45.55%, respectively. Due to the properties of controllable stiffness and damping of MRE, the isolation transmissibility and root mean square of acceleration response can be reduced by 41.2% and 65.3%, respectively. The proposed MRE isolator can be used as a controllable stiffness device and has great potential in the field of vibration suppression for heavy equipment. [116].

4.3 Other applications

In addition to the field-sensitive elastic property, MREs possess several functions such as magnetoelasticity, magnetoresistance, magnetostriction, piezoresistance, and thermoresistance [1, 119]. The reasons for these functions are the changes in the spacing between the magnetic particles due to external loadings, which produce variations of the physical or mechanical properties of the MRE materials. Based on their field-sensitive properties, MREs have been developed for use as sensors and actuators, for example, force sensor [120], magnetoresistive sensor [18], magneto-sensitive strain sensor [20], flexible tri-axis tactile sensor [21], self-powered tribo-sensor [22], combined magnetic and mechanical sensor [17, 121], soft actuator [122], actuators for valves [123], MEMS magnetometer [124], etc. Moreover, the microwave response [125, 126] and 3D printing properties of MREs [127–129] have also been recently reported. It is worth being pointed out that the application of MREs is explosively developing.

5. Summary

In this chapter, the materials and applications of MREs are briefly reviewed. Firstly, raw materials, including the magnetic particles, the rubber-like matrices, and the additives, are introduced. As the kind of the raw materials is getting more and more inconstant, there are a variety of raw materials that can be used to prepare MREs. Attributing to the variety of the raw materials, the kinds of prepared MREs are much various and the study on the MRE materials is a long-lasting discovery or inventive subject. Meanwhile, the XμCT technique can be used to study the microstructures and microstructure-based mechanisms of MREs. With the development of MRE materials, the property of MREs gets more and more various. The application of MREs has quickly developed in the engineering field of vibration reduction and the application range of MREs is getting wider and wider. In addition to the conventional applications in vibration reduction, the sensing and actuating applications of MREs are recently explosively developed. Moreover, the microwave-response and the 3D printing of MREs are newly emerged subjects, which can be much promising in engineering applications in the near future.

Acknowledgements

The support of the National Natural Science Foundation of China (Grant Nos. 11602242 and 11502256) is acknowledged. The authors are grateful for the helpful suggestions of Prof. Xinglong Gong at the University of Science and Technology of China and the support of the colleagues who provided the original figures for the article.

Conflict of interest

There is no conflict of interest.

Author details

Taixiang Liu[1] and Yangguang Xu[2*]

1 Research Center of Laser Fusion, China Academy of Engineering Physics, Mianyang, China

2 College of Aerospace Engineering, Chongqing University, Chongqing, China

*Address all correspondence to: xyg@mail.ustc.edu.cn

References

[1] Li YC, Li JC, Li WH, Du HP. A state-of-the-art review on magnetorheological elastomer devices. Smart Materials and Structures. 2014;**23**:123001. DOI: 10.1088/0964-1726/23/12/123001

[2] Ginder JM, Nichols ME, Dlie LD, Tardiff JL. Magnetorheological elastomers: Properties and applications. Proceedings of SPIE. 1999;**3675**:131-138. DOI: 10.1117/12.352787

[3] Carlson JD, Jolly MR. MR fluid, foam and elastomer devices. Mechatronics. 2000;**10**:555-569. DOI: 10.1016/s0957-4158(99)00064-1

[4] Gong XL, Zhang XZ, Zhang PQ. Fabrication and characterization of isotropic magnetorheological elastomers. Polymer Testing. 2005;**24**:669-676. DOI: 10.1016/j.polymertesting.2005.03.015

[5] Li WH, Zhang XZ. Research and applications of MR elastomers. Recent Patents on Engineering. 2008;**1**:161-166. DOI: 10.2174/1874477X10801030161

[6] Rigbi Z, Jilkén L. The response of an elastomer filled with soft ferrite to mechanical and magnetic influences. Journal of Magnetism and Magnetic Materials. 1983;**37**:267-276. DOI: 10.1016/0304-8853(83)90055-0

[7] Rabinow J. The magnetic fluid cluth. AIEE Transactions. 1948;**67**:1308-1315. DOI: 10.1109/t-aiee.1948.5059821

[8] Ashour O, Rogers CA, Kordonsky W. Magnetorheological fluids: Materials, characterization, and devices. Journal of Intelligent Material Systems and Structures. 1996;**7**:123-130. DOI: 10.1177/1045389X9600700201

[9] Bossis G, Volkova O, Lacis S, Meunier A. Magnetorheology: fluids, structures and rheology. In: Odenbach S, editor. Ferrofluids, Lecture Notes in Physics. Heidelberg: Springer; 2002. pp. 202-230. DOI: 10.1007/3-540-45646-5_11

[10] de Vicente J, Klingenberg DJ, Hidalgo-Alvarez R. Magnetorheological fluids: A review. Soft Matter. 2011;**7**:3701-3710. DOI: 10.1039/c0sm01221a

[11] Choi HJ, Kim CA, Kwon TM, Jhon MS. Viscosity of magnetic particle suspensions. Journal of Magnetism and Magnetic Materials. 2000;**209**:228-230. DOI: 10.1016/s0304-8853(99)00695-2

[12] López-López MT, Zugaldía A, González-Caballero F, Durán JDG. Sedimentation and redispersion phenomena in iron-based magnetorheological fluids. Journal of Rheology. 2006;**50**:543-560. DOI: 10.1122/1.2206716

[13] Deng HX, Gong XL. Application of magnetorheological elastomers to vibration absorber. Communications in Nonlinear Science and Numerical Simulation. 2008;**13**:1938-1947. DOI: 10.1016/j.cnsns.2007.03.024

[14] Popp KM, Kroger M, Li WH, Zhang XZ, Kosasih P. MRE properties under shear and squeeze modes and applications. Journal of Intelligent Material Systems and Structures. 2010;**21**:1471-1477. DOI: 10.1177/1045389X09355666

[15] Li WH, Zhang XZ, Du H. Magnetorheological elastomers and their applicatioins. In: Visakh PM, Thomas S, Chandra AK, Mathew AP, editors. Advances in Elastomers I: Blends and Interpenetrating Networks. Berlin: Springer; 2013. pp. 357-374. DOI: 10.1007/978-3-642-20925-3_12

[16] Wang Q, Dong XF, Li LY, Ou JP. Study on an improved variable stiffness tuned mass damper based on conical magnetorheological

elastomer isolators. Smart Materials and Structrues. 2017;**26**:105028. DOI: 10.1088/1361-665X/aa81e8

[17] Wang XJ, Gordaninejad F, Calgar M, Liu YM. Sensing behavior of magnetorheological elastomers. Journal of Mechanical Design. 2009;**131**:091004. DOI: 10.1115/1.3160316

[18] Bica I. Magnetoresistor sensor with magnetorheological elastomers. Journal of Industrial and Engineering Chemistry. 2011;**17**:83-89. DOI: 10.1016/j.jiec.2010.12.001

[19] Ausanio G, Iannotti V, Ricciardi E, Lanotte L, Lanotte L. Magneto-piezoresistance in magnetorheological elastomers for magnetic induction gradient or position sensors. Sensors and Actuators A: Physical. 2014;**205**:235-239. DOI: 10.1016/j.sna.2013.10.009

[20] Hu T, Xuan SH, Ding L, Gong XL. Stretchable and magneto-sensitive strain sensor based on silver nanowire-polyurethane sponge enhanced magnetorheological elastomer. Materials and Design. 2018;**156**:528-537. DOI: 10.1016/j.matdes.2018.07.024

[21] Kawasetsu T, Horii T, Ishihara H, Asada M. Flexible tri-axis tactile sensor using spiral inductor and magnetorheological elastomer. IEEE Sensors Journal. 2018;**18**:5834-5841. DOI: 10.1109/JSEN.2018.2844194

[22] Qi S, Guo HY, Chen J, Hu CG, Yu M, Wang ZL. Magnetorheological elastomers enabled high-sensitive self-powered tribo-sensor for magnetic field detection. Nanscale. 2018;**10**:4745-4752. DOI: 10.1039/c7nr09129j

[23] Japka JE. Microstructure and properties of carbonyl iron powder. Journal of Metals. 1988;**40**:18-21. DOI: 10.1007/BF03258115

[24] Carbonyl iron [Internet]. Available from: https://en.wikipedia.org/wiki/Carbonyl_iron [Accessed: Dec 30, 2018]

[25] Chen L, Gong XL, Jiang WQ, Yao JJ, Deng HX, Li WH. Investigation on magnetorheological elastomers based on natural rubber. Journal of Materials Science. 2007;**42**:5483-5489. DOI: 10.1007/s10853-006-0975-x

[26] Khimi SR, Pickering KL, Mace BR. Dynamic properties of magnetorheological elastomers based on iron sand and natural rubber. Journal of Applied Polymer Science. 2015;**132**:41506. DOI: 10.1002/app.41506

[27] Bose H. Viscoelastic properties of silicone-based magnetorheological elastomers. International Journal of Modern Physics B. 2007;**21**:4790-4797. DOI: 10.1142/S0217979207045670

[28] Zhang W, Gong XL, Li JF, Zhu H, Jiang WQ. Radiation vulcanization of magnetorheological elastomers based on silicone rubber. Chinese Journal of Chemical Physics. 2009;**22**:535-540. DOI: 10.1088/1674-0068/22/05/535-540

[29] Li R, Sun LZ. Viscoelastic responses of silicone-rubber-based magnetorheological elastomers under compressive and shear loadings. Journal of Engineering Materials and Technology. 2013;**135**:021008. DOI: 10.1115/1.4023839

[30] Bunoiu M, Bica I. Magnetorheological elastomer based on silicone rubber, carbonyl iron and Rochelle salt: Effects of alternating electric and static magnetic fields intensities. Journal of Industrial and Engineering Chemistry. 2016;**37**:312-318. DOI: 10.1016/j.jiec.2016.03.047

[31] Japka JE. Silicone rubber based magnetorheological elastomer: magnetic structure tested by means of neutron depolarization and magnetic force microscopy methods. Journal of Physics:

Conference Series. 2017;**848**:012016. DOI: 10.1088/1742-6596/848/1/012016

[32] Tian TF, Zhang XZ, Li WH, Alici G, Ding J. Study of PDMS based magnetorheological elastomers. Journal of Physics: Conference Series. 2013;**412**:012038. DOI: 10.1088/1742-6596/412/1/012038

[33] Li WH, Nakano M. Fabrication and characterization of PDMS based magnetorheological elastomers. Smart Materials and Structures. 2013;**22**:055035. DOI: 10.1088/0964-1726/22/5/055035

[34] Perales-Martínez IA, Palacios-Pineda LM, Lozano-Sanchez LM, Martínez-Romero O, Puente-Cordova JG, Elías-Zúniga A. Enhancement of a magnetorheological PDMS elastomer with carbonyl iron particles. Polymer Testing. 2017;**57**:78-86. DOI: 10.1016/j.polymertesting.2016.10.029

[35] Danas K, Kankanala SV, Triantafyllidis N. Experiments and modeling of iron-particle-filled of magnetorheological elastomers. Journal of the Mechanics and Physics of Solids. 2012;**60**:120-138. DOI: 10.1016/j.jmps.2011.09.006

[36] Zhang W, Gong XL, Xuan SH, Jiang WQ. Temperature-dependent mechanical properties and model of magnetorheological elastomers. Industrial & Engineering Chemistry Research. 2011;**50**:6704-6712. DOI: 10.1021/ie200386x

[37] Gong XL, Fan YC, Xuan SH, Xu YG, Peng C. Control of the damping properties of magnetorheological elastomers by using polycaprolactone as a temperature-controlling component. Industrial & Engineering Chemistry Research. 2012;**51**:6395-6403. DOI: 10.1021/ie300317b

[38] Ju BX, Tang R, Zhang DY, Yang BL, Yu M, Liao CR. Temperature-dependent dynamic mechanical properties of magnetorheological elastomers under magnetic field. Journal of Magnetism and Magnetic Materials. 2015;**374**:283-288. DOI: 10.1016/j.jmmm.2014.08.012

[39] Xiang CL, Gao P, Liu H, Zhou H. Experimental and theoretical study of temperature-dependent variable stiffness of magnetorheological elastomers. International Journal of Materials Research. 2018;**109**:1-16. DOI: 10.3139/146. 111590

[40] Lian CL, Lee KH, Lee CH. Effect of temperature and relative humidity on friction and wear properties of silicone-based magnetorheological elastomer. Tribology Transactions. 2018;**61**:238-246. DOI: 10.1080/10402004.2017.1306636

[41] Liao GJ, Xu YG, Wang FJ, Wei FY, Wan Q. Influence of γ radiation on the shear modulus of magnetorheological elastomer. Materials Letters. 2016;**174**:79-81. DOI: 10.1016/j.matlet.2016.03.085

[42] Ginder JM, Nichols ME, Elie LD, Clark SM. Controllable-stiffness components based on magnetorheological elastomers. Proceedings of SPIE. 2000;**3985**: 418-425. DOI: 10.1117/12.388844

[43] Ubaidillah SJ, Purwanto A, Mazlan SA. Recent Progress on Magnetorheological solids: materials, fabrication, testing, and applications. Advanced Engineering Materials. 2015;**17**:563-597. DOI: 10.1002/adem.201400258

[44] BASF SE [Internet]. Available from: http://www.monomers.basf.com [Accessed: Dec 30, 2018]

[45] Zhu H. Fabrication of practical magnetorheological fluids and their properties [thesis]. Hefei: University of Science and Technology of China; 2010

[46] Liu TX, Gong XL, Xu YG, Xuan SH, Jiang WQ. Simulation of magneto-induced rearrangeable microstructures of magnetorheological plastomers. Soft Matter. 2013;**9**:10069-10080. DOI: 10.1039/c3sm52130c

[47] Lokander M, Stenberg B. Performance of isotropic magnetorheological rubber materials. Polymer Testing. 2003;**22**:245-251. DOI: 10.1016/S0142-9418(02)00043-0

[48] Stepanov GV, Abramchuk SS, Grishin DA, Nikitin LV, Kramarenko EY, Khokhlov AR. Effect of a homogeneous magnetic field on the viscoelastic behavior of magnetic elastomers. Polymer. 2007;**48**:488-495. DOI: 10.1016/ j.polymer.2006.11.044

[49] Liu TX. Study on the magneto-mechanical behavior of magnetorheological plastomers and its microstructure-based mechanism [thesis]. Hefei: University of Science and Technology of China; 2015

[50] Silicone rubber [Internet]. Available from: https://en.wikipedia.org/wiki/Silicone_rubber [Accessed: Dec 30, 2018]

[51] Natural rubber [Internet]. Available from: https://en.wikipedia.org/wiki/Natural_rubber [Accessed: Dec 30, 2018]

[52] Polybutadiene [Internet]. Available from: https://en.wikipedia.org/wiki/Polybutadiene [Accessed: Dec 30, 2018]

[53] Butyl_rubber [Internet]. Available from: https://en.wikipedia.org/wiki/Butyl_rubber [Accessed: Dec 30, 2018]

[54] Polyurethane [Internet]. Available from: https://en.wikipedia.org/wiki/Polyurethane [Accessed: Dec 30, 2018]

[55] Polydimethylsiloxane [Internet]. Available from: https://en.wikipedia.org/wiki/Polydimethylsiloxane [Accessed: Dec 30, 2018]

[56] Epoxy [Internet]. Available from: https://en.wikipedia.org/wiki/Epoxy [Accessed: Dec 30, 2018]

[57] Characteristic properties of silicone rubber compounds [Internet]. Available from: http://www.shinetsusilicone-global.com/ [Accessed: Dec 30, 2018]

[58] Leblanc J. Rubber-filler interactions and rheological properties in filled compounds. Progress in Polymer Science. 2002;**27**:627-687. DOI: 10.1016/s0079-6700(01)00040-5

[59] Chen L, Gong XL, Li WH. Effect of carbon black on the mechanical performances of magnetorheological elastomers. Polymer Testing. 2008;**27**:340-345. DOI: 10.1016/j.polymertesting.2007.12.003

[60] Nayak B, Dwivedy S, Murthy K. Fabrication and characterization of magnetorheological elastomer with carbon black. Journal of Intelligent Material Systems and Structures. 2015;**26**:830-839. DOI: 10.1177/1045389X14535011

[61] Lu HL, Wang WJ, Yang FF, Wang GP, Rui XT. Effect of carbon black with large particle size on dynamic mechanical analysis of magnetorheological elastomers (MREs). Materials Research Express. 2018;**5**:095703. DOI: 10.1088/2053-1591/aad88b

[62] Aziz S, Mazlan S, Ismail N, Ubaidillah U, Choi S, Khairi M, et al. Effects of multiwall carbon nanotubes on viscoelastic properties of magnetorheological elastomers. Smart Materials and Structures. 2016;**25**:077001. DOI: 10.1088/0964-1726/25/7/077001

[63] Aziz S, Mazlan S, Ismail N, Ubaidillah U, Khairi M, Yunus N. Rheological properties of carbon nanotubes-reinforced magnetorheological elastomer.

Journal of Physics: Conference Series. 2017;**795**:012074. DOI: 10.1088/1742-6596/795/1/012074

[64] Poojary U, Hegde S, Gangadharan K. Experimental investigation on the effect of carbon nanotube additive on the field-induced viscoelastic properties of magnetorheological elastomer. Journal of Materials Science. 2018;**53**:4229-4241. DOI: 10.1007/s10853-017-1883-y

[65] Aziz S, Ubaidillah U, Mazlan S, Ismail N, Choi SB. Implementation of functionalized multiwall carbon nanotubes on magnetorheological elastomer. Journal of Materials Science. 2018;**53**:10122-10134. DOI: 10.1007/s10853-018-2315-3

[66] Lee CJ, Kwon SH, Choi HJ, Chung KH, Jung JH. Enhanced magnetorheological performance of carbonyl iron/natural rubber composite elastomer with gamma-ferrite additive. Colloid and Polymer Science. 2018;**296**:1609-1613. DOI: 10.1007/s00396-018-4373-0

[67] Gordaninejad F, Wang XJ, Mysore P. Behavior of thick magnetorheological elastomers. Journal of Intelligent Material Systems and Structures. 2012;**23**:1033-1039. DOI: 10.1177/1045389X12448286

[68] Chen L, Gong XL, Li WH. Microstructures and viscoelastic properties of anisotropic magnetorheological elastomers. Smart Materials and Structures. 2007;**16**:2645-2650. DOI: 10.1088/0964-1726/16/6/069

[69] Bakshi SR, Batista RG, Agarwal A. Quantification of carbon nanotube distribution and property correlation in nanocomposites. Composites: Part A. 2009;**40**:1311-1318. DOI: 10.1016/j.compositesa.2009.06.004

[70] Balasoiu M, Craus ML, Anitas EM, Bica I, Plestil J, Kuklin AI. Microstructure of stomaflex based magnetic elastomers. Physics of the Solid State. 2010;**52**:917-921. DOI: 10.1134/s1063783410050070

[71] Gunther D, Borin D, Gunther S, Odenbach S. X-ray micro-tomographic characterization of field-structured magnetorheological elastomers. Smart Materials and Structures. 2012;**21**:015005. DOI: 10.1088/0964-1726/21/1/015005

[72] Borbath T, Gunther S, Borin D, Gundermann T, Odenbach S. XμCT analysis of magnetic field-induced phase transitions in magnetorheological elastomers. Smart Materials and Structures. 2012;**21**:105018. DOI: 10.1088/0964-1726/21/10/105018

[73] Gundermann T, Odenbach S. Investigation of the motion of particles in magnetorheological elastomers by X-μCT. Smart Materials and Structures. 2014;**23**:105013. DOI: 10.1088/0964-1726/23/10/105013

[74] Jolly MR, Carlson JD, Munoz B. A model of the behaviour of magnetorheological materials. Smart Materials and Structures. 1996;**5**:607-614. DOI: 10.1088/0964-1726/5/5/009

[75] Davis LC. Model of magnetorheological elastomers. Journal of Applied Physics. 1999;**85**:3348-3351. DOI: 10.1063/1.369682

[76] Li WH, Zhang XZ. A study of the magnetorheological effect of bimodal particle based magnetorheological elastomers. Smart Materials and Structures. 2010;**19**:035002. DOI: 10.1088/0964-1726/19/3/035002

[77] Metsch P, Kalina KA, Spieler C, Kastner M. A numerical study on magnetostrictive phenomena in magnetorheological elastomers. Computational Materials Science. 2016;**124**:364-374. DOI: 10.1016/j.commatsci.2016.08.012

[78] Kalina KA, Metsch P, Kastner M. Microscale modeling and simulation of magnetorheological elastomers at finite strains: A study on the influence of mechanical preloads. International Journal of Solids and Structures. 2016;**102-103**:286-296. DOI: 10.1016/j.ijsolstr.2016.10.019

[79] Liu TX, Xu YG. Soft magneto-sensitive particle-reinforced composite material and its tunable three-dimensional microstructure. In: The 18th U.S. National Congress for Theoretical and Applied Mechanics; 5-9 June 2018; Chicago, USA

[80] Shen Y, Golnaraghi MF, Heppler GR. Experimental research and modeling of magnetorheological elastomers. Journal of Intelligent Material Systems and Structures. 2004;**15**:27-35. DOI: 10.1177/1045389X04039264

[81] Biller AM, Stolbov OV, Raikher YL. Modeling of particle interactions in magnetorheological elastomers. Journal of Applied Physics. 2014;**116**:114904. DOI: 10.1063/1.4895980

[82] Biller AM, Stolbov OV, Raikher YL. Two-particle element of a magnetorheological elastomer under a cyclic magnetic field. Journal of Physics: Conference Series. 2018;**994**:012001. DOI: 10.1088/1742-6596/994/1/012001

[83] Dorfmann A, Ogden RW. Magnetoelastic modelling of elastomers. European Journal of Mechanics A/Solids. 2003;**22**:497-507. DOI: 10.1016/S0997-7538(03)00067-6

[84] Chatzigeorgiou G, Javili A, Steinmann P. Unified magnetomechanical homogenization framework with application to magnetorheological elastomers. Mathematics and Mechanics of Solids. 2014;**19**:193-211. DOI: 10.1177/1081286512458109

[85] Ethiraj G, Sridhar A, Miehe C. A Magneto-visco-elastic model for magnetorheological elastomers. Proceedings of Applied Mathematics and Mechanics. 2014;**14**:515-516. DOI: 10.1002/pamm.201410245

[86] Agirre-Olabide I, Lion A, Elejabarrieta MJ. A new three-dimensional magneto-viscoelastic model for isotropic magnetorheological elastomers. Smart Materials and Structures. 2017;**26**:035021. DOI: 10.1088/1361-665X/26/3/035021

[87] Kou Y, Jin K, Xu LQ, Zheng XJ. A novel phenomenological model for dynamic behavior of magnetorheological elastomers in tension–compression mode. Smart Materials and Structures. 2017;**26**:065011. DOI: 10.1088/1361-665X/aa6126

[88] Kou Y, Jin K, Xu LQ, Zheng XJ. A visoelastic constitutive model for magneto-mechanical coupling of magnetorheological elastomers. Smart Materials and Structures. 2017;**26**:115017. DOI: 10.1088/1361-665X/aa8d3d

[89] Cantera MA, Behrooz M, Gibson RF, Gordaninejad F. Modeling of magneto-mechanical response of magnetorheological elastomers (MRE) and MRE-based systems: a review. Smart Materials and Structures. 2017;**26**:023001. DOI: 10.1088/1361-665X/aa549c

[90] Kalina KA, Brummund J, Metsch P, Kastner M. Microscale modeling and simulation of magnetorheological elastomers. Proceedings of Applied Mathematics and Mechanics. 2017;**17**:27-30. DOI: 10.1002/pamm.201710008

[91] Wang Q, Dong XF, Li LY, Ou JP. A nonlinear model of magnetorheological elastomer with wide amplitude range and variable

frequencies. Smart Materials and Structures. 2017;**26**:065010. DOI: 10.1088/1361-665X/aa66e3

[92] Agirre-Olabide KP, Elejabarrieta MJ. Linear magneto-viscoelastic model based on magnetic permeability components for anisotropic magnetorheological elastomers. Journal of Magnetism and Magnetic Materials. 2018;**446**:155-161. DOI: 10.1016/j.jmmm.2017.09.017

[93] Wang Q, Dong XF, Li LY, Ou JP. Mechanical modeling for magnetorheological elastomer isolators based on constitutive equations and electromagnetic analysis. Smart Materials and Structures. 2018;**27**:065017. DOI: 10.1088/1361-665X/aabdb5

[94] Li WH, Zhang XZ, Du HP. Development and simulation evaluation of a magnetorheological elastomer isolator for seat vibration control. Journal of Intelligent Material Systems and Structures. 2012;**23**:1041-1048. DOI: 10.1177/1045389X11435431

[95] Ginder JM, Clark SM, Schlotter WF, Nichols ME. Magnetostrictive phenomena in magnetorheological elastomer. International Journal of Modern Physics B. 2002;**16**:2412-2418. DOI: 10.1142/S021797920201244X

[96] Guan XC, Dong XF, Ou JP. Magnetostrictive effect of magnetorheological elastomer. Journal of Magnetism and Magnetic Materials. 2008;**320**:158-163. DOI: 10.1016/j.jmmm.2007.05.043

[97] Sun SL, Peng XQ, Guo ZY. Nonlinear magnetostrictive effect of magnetorheological elastomers. Advanced Materials Research. 2014;**833**:291-294. DOI: 10.4028/www.scientific.net/AMR.833.291

[98] Gong XL, Liao GJ, Xuan SH. Full-field deformation of magnetorheological elastomer under uniform magnetic field. Applied Physics Letters. 2012;**100**:211909. DOI: 10.1063/1.4722789

[99] Kchit N, Bossis G. Electrical resistivity mechanism in magnetorheological elastomer. Journal of Physics D: Applied Physics. 2009;**42**:105505. DOI: 10.1088/0022-3727/42/10/105505

[100] Zhu XL, Meng YG, Tian Y. Nonlinear pressure-dependent conductivity of magnetorheological elastomers. Smart Materials and Structures. 2010;**19**:117001. DOI: 10.1088/0964-1726/19/11/117001

[101] Bica I. The influence of hydrostatic pressure and transverse magnetic field on the electric conductivity of the magnetorheological elastomers. Journal of Industrial and Engineering Chemistry. 2012;**18**:483-486. DOI: 10.1016/j.jiec.2011.11.067

[102] Bica I, Anitas EM, Bunoiu M, Vatzulik B, Juganaru I. Hybrid magnetorheological elastomer: Influence of magnetic field and compression pressure on its electrical conductivity. Journal of Industrial and Engineering Chemistry. 2014;**20**:3994-3999. DOI: 10.1016/j.jiec.2013.12.102

[103] Ge L, Gong XL, Wang Y, Xuan SH. The conductive three dimensional topological structure enhanced magnetorheological elastomer towards a strain sensor. Composites Science and Technology. 2016;**135**:92-99. DOI: 10.1016/j.compscitech.2016.09.015

[104] Wang Y, Xuan SH, Dong B, Xu F, Gong XL. Stimuli dependent impedance of conductive magnetorheological elastomers. Smart Materials and Structures. 2016;**25**:025003. DOI: 10.1088/0964-1726/25/2/025003

[105] Wang Y, Xuan SH, Ge L, Wen QQ, Gong XL. Conductive

magnetorheological elastomer: Fatigue dependent impedance-mechanic coupling properties. Smart Materials and Structures. 2017;**26**:015004. DOI: 10.1088/0964-1726/26/1/015004

[106] Kordonsky WI. Magnetorheological effect as a base of new devices and technologies. Journal of Magnetism and Magnetic Materials. 1993;**122**:395-398. DOI: 10.1016/0304-8853(93)91117-P

[107] Ginder JM, Schlotter WF, Nichols ME. Magnetorheological elastomers in tunable vibration absorbers. Proceedings of SPIE. 2001;**4331**:103-110. DOI: 10.1117/12.432694

[108] Deng HX, Gong XL, Wang LH. Development of an adaptive tuned vibration absorber with magnetorheological elastomer. Smart Materials and Structures. 2006;**15**:N111-N116. DOI: 10.1088/0964-1726/15/5/N02

[109] Hoang N, Zhang N, Du H. An adaptive tunable vibration absorber using a new magnetorheological elastomer for vehicular powertrain transient vibration reduction. Smart Materials and Structures. 2011;**20**:015019. DOI: 10.1088/0964-1726/20/1/015019

[110] Liao GJ, Gong XL, Xuan SH. Phase based stiffness tuning algorithm for a magnetorheological elastomer dynamic vibration absorber. Smart Materials and Structures. 2014;**23**:015016. DOI: 10.1088/09641726/23/1/015016

[111] Xin FL, Bai XX, Qian LJ. Principle, modeling, and control of a magnetorheological elastomer dynamic vibration absorber for powertrain mount systems of automobiles. Journal of Intelligent Material Systems and Structures. 2017;**28**:2239-2254. DOI: 10.1177/1045389X16672731

[112] Sun SS, Yang J, Yildirim T, Du HP, Alici G, Zhang SW, et al. Development of a nonlinear adaptive absorber based on magnetorheological elastomer. Journal of Intelligent Material Systems and Structures. 2018;**29**:194-204. DOI: 10.1177/1045389X17733053

[113] Ibrahim RA. Recent advances in nonlinear passive vibration isolators. Journal of Sound and Vibration. 2008;**314**:371-452. DOI: 10.1016/j.jsv.2008.01.014

[114] Yang J, Du HP, Li WH, Li YC, Li JC, Sun SS, et al. Experimental study and modeling of a novel magnetorheological elastomer isolator. Smart Materials and Structures. 2013;**22**:117001. DOI: 10.1088/0964-1726/22/11/117001

[115] Behrooz M, Wang XJ, Gordaninejad F. Modeling of a new semi-active/passive magnetorheological elastomer isolator. Smart Materials and Structures. 2014;**23**:045003. DOI: 10.1088/0964-1726/23/4/045013

[116] Yu M, Zhao LJ, Fu J, Zhu M. Thermal effects on the laminated magnetorheological elastomer isolator. Smart Materials and Structures. 2016;**25**:115039. DOI: 10.1088/0964-1726/25/11/115039

[117] Li YC, Li JC. On rate-dependent mechanical model for adaptive magnetorheological elastomer base isolator. Smart Materials and Structures. 2017;**26**:045001. DOI: 10.1088/1361-665X/aa5f95

[118] Tao Y, Rui XT, Yang FF, Chen GL, Bian LX, Zhu W, et al. Design and experimental research of a magnetorheological elastomer isolator working in squeeze/elongation-shear mode. Journal of Intelligent Material Systems and Structures. 2018;**29**:1418-1429. DOI: 10.1177/1045389X17740436

[119] Martin JE, Anderson RA, Odinek J, Adolf D, Williamson J. Controlling percolation in field-structured particle composites: Observations of giant

thermoresistance, piezoresistance, and chemiresistance. Physical Review B. 2003;**67**:094207. DOI: 10.1103/PhysRevB.67.094207

[120] Li WH, Kostidis K, Zhang XZ, Zhou Y. Development of a force sensor working with MR elastomers. 2009 IEEE/ASME International Conference on Advanced Intelligent Mechatronics; 14-17 July 2009; Singapore

[121] Ghafoorianfar N, Wang XJ, Gordaninejad F. Combined magnetic and mechanical sensing of magnetorheological elastomers. Smart Materials and Structures. 2014;**23**:055010. DOI: 10.1088/0964-1726/23/5/055010

[122] Kashima S, Miyasaka F, Hirata K. Novel soft actuator using magnetorheological elastomer. IEEE Transactions on magnetics. 2012;**48**:1649-1652. DOI: 10.1109/TMAG.2011.2173669

[123] Bose H, Rabindranath R, Ehrlich J. Soft magnetorheological elastomers as new actuators for valves. Journal of Intelligent Material Systems and Structures. 2011;**23**:989. DOI: 10.1177/1045389X11433498

[124] Du GT, Chen XD. MEMS magnetometer based on magnetorheological elastomer. Measurement. 2012;**45**:54-58. DOI: 10.1016/j.measurement.2011.10.002

[125] Butera A, Alvarez N, Jorge G, Ruiz MM, Mietta JL, Negri RM. Microwave response of anisotropic magnetorheological elastomers: Model and experiments. Physical Review B. 2012;**86**:144424. DOI: 10.1103/PhysRevB.86.144424

[126] Yu M, Yang PA, Fu J, Liu SZ, Qi S. Study on the characteristics of magnetosensitive electromagnetic wave-absorbing properties of magnetorheological elastomers. Smart Materials and Structures. 2016;**25**:085046. DOI: 10.1088/0964-1726/25/8/085046

[127] Bastola AK, Hoang VT, Li L. A novel hybrid magnetorheological elastomer developed by 3D printing. Materials and Design. 2017;**114**:391-397. DOI: 10.1016/j.matdes.2016.11.006

[128] Bastola AK, Paudel M, Li L. Patterned magnetorheological elastomer developed by 3D printing. Materials Science Forum. 2018;**939**:147-152. DOI: 10.4028/www.scientific.net/MSF.939.147

[129] Bastola AK, Paudel M, Li L. Development of hybrid magnetorheological elastomers by 3D printing. Polymer. 2018;**149**:213-228. DOI: 10.1016/j.polymer.2018.06.076

Chapter 5

Dual Response of Materials under Electric and Magnetic Fields

Mehmet Cabuk

Abstract

The electrorheological (ER) effect is known as the change in the rheological behaviors of ER fluids under applied electric field E. When an E is imposed, ER fluids show phase transition from a liquid to a solid-like state due to the interactions of polarized particles. This solid-like behavior of particles is due to the increasing viscosity of suspensions. ER materials belong to a family of controllable fluids. ER fluids are dispersions of solid particles in a hydrophobic insulating dispersion medium. These solid particles play a very important role in the ER activity of dispersions. As the dispersed phase, diverse materials such as polymer blends, gels, biodegradable materials, clays, graphene oxide, hybrid nanocomposites, copolymers, ionic liquids, and conducting polymers have been proposed. In the magnetorheological fluids, this control is provided with magnetic field. Various magnetic particles such as carbonyl iron and iron oxides have been suggested as MR material. The combined effect of magnetic and electric field produces intensified rheological changes in the suspensions. This synergic effect is termed as electromagnetorheological effect (EMR). The EMR effect provides a new strategy to control the rheological properties of dispersions.

Keywords: electrorheology, magnetorheology, nanocomposites, smart fluids, dual response

1. Introduction

The electrorheological (ER) effect is known as the change in the rheological behaviors of ER fluids under applied electric field E. When an E is imposed, ER fluids show phase transition from a liquid to a solid-like state due to the interactions of polarized particles. This solid-like behavior of particles is due to the increasing viscosity of suspensions. There are two main driving forces behind the viscosity increase in ER fluids. Fibrillar structure of polarized molecules with applied electric field and ion aggregation near the electrode surfaces are responsible for the viscosity increase. Essentially the ER effect in a liquid is related to the motions of ions or polar molecules [1, 2] (**Figure 1**).

In the nineteenth century, the apparent viscosity of pure nonconducting liquids was believed to increase with the applied E. This effect was called as electroviscous effect rather than the ER effect [3]. Many other solution systems showed that the electrolyte solutions have very much stronger electroviscous effect than the pure liquid. After this time, Winslow studied the E-induced viscosity increase of solid semiconducting particles dispersed in a low viscosity and high insulating oil medium [4]. The E-induced effect was much stronger than the electroviscous effect.

Therefore, the ER effect was termed as "Winslow effect." Winslow effect is regarded as the formation of fibrillated chain structures by the dispersed particles between the electrodes of a measuring system. Efforts have been carried out to elucidate the ER mechanism and to improve ER performance of dispersions that meets the industrial requirements.

ER fluids exhibit large reversible changes in their rheological properties when they are subjected to external E. Many studies have been carried out on the mechanism and application of ER fluid. One important property of the ER fluids is that it shows quick response to electric field. This inherent behavior has pioneered many researches to various engineering applications [5]. Some of these are machine mounts, shock absorbers, and smart structures. The rheological properties of ER fluids can be electrically controlled, and this control is beneficial to a wide range of application technologies such as active vibration, damping, and resistive force generation. Several commercial applications have been developed until today, and many potential applications especially in the automotive industry such as ER fluid-based engine mount, shock absorber, clutches, and seat dampers are still undiscovered [6].

Many materials have been used for the preparation of ER fluids. They can be classified as inorganic materials [7, 8] (such as TiO_2, $BaTiO_2$), organic materials [9] (such as chitosan, alginate, and its derivatives), and conducting polymers (such as polyaniline, polypyrrole). Among them conducting polymers have been widely studied, because of their ease of synthesis, less corrosive properties and high stability, and compatibility with the carrier liquid.

Yield shear stress is one of the most important parameters for ER fluids and can be varied by the applied E. Despite these properties, ER fluids have limited their application due to low yield stress, which yields poor mechanical performance that limits their application [1, 10]. The developing of the giant ER effect has revived interest in this area. When compared with traditional ER fluids, giant ER fluids show higher magnitude yield stresses. The polarization force of molecular dipoles between dispersed particles is responsible for the giant ER effect [11]. Many organic–/inorganic-structured giant ER materials have been synthesized. Wen et al. [12] reported the synthesis and ER properties of a new type of giant ER fluid consisting of polar group-modified nano-sized barium titanyl oxalate particles. Also, the problem associated with sedimentation of colloidal particles, as well as the lack of high-performance materials, has inhibited broad engineering applications [13, 14]. Giant ER fluids like other traditional ER fluids exhibit sedimentation drawback due to the density mismatch of the fluid and solid phases as well as the aggregation of the dispersed particles [15, 16]. To improve the sedimentation stability of the ER

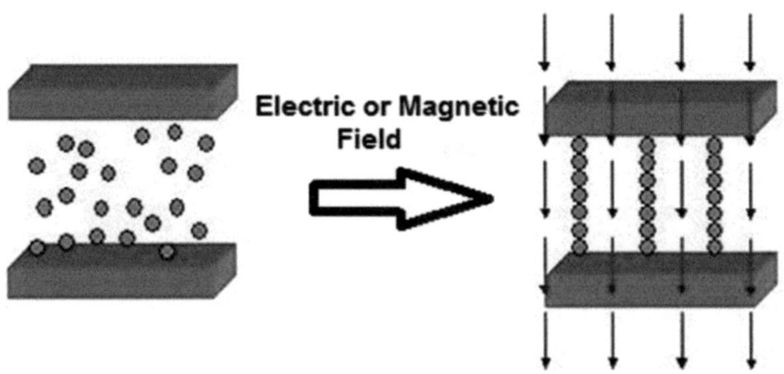

Figure 1.
Schematic illustration of particles into an ER fluid at absence and presence of electric or magnetic field.

fluids, several methods have been developed. They can be listed as follows by adding surfactant to the dispersion, making the particle phase less dense that decrease the density mismatch and modify the surface or particle morphology [17]. Li et al. [18] reported the fabrication of multiwall carbon nanotubes, urea-coated barium titanyl oxylate composite particles, that exhibit the giant ER effect (yield stress was 194 kPa at 5 kV/mm). The antisedimentation property enhanced dramatically in silicone oil due to multiwall carbon nanotubes. Cabuk et al. [16] reported ER response and temperature effect on antisedimentation stability of expanded perlite particles in silicone oil. The expanded perlite particles as porous ultra-lightweight materials showed typical ER properties under applied E, exhibiting a shear thinning non-Newtonian viscoelastic behavior (**Figure 2**).

When the mechanism of the ER fluids is examined, Most ER fluids fit the Bingham fluid model. At this model, an ER fluid has its own yield stress, and the flow motion of the fluid impedes when an applied external shear stress is lower than the yield stress [19]. The Bingham model has been extensively used as a viscoelastic equation:

$$\tau = \tau_y + \eta\dot{\gamma}, \tau \geq \tau_y \tag{1}$$

$$\dot{\gamma} = 0, \tau < \tau_y \tag{2}$$

In Eqs. (1) and (2), τ and τ_y are shear and yield stresses, $\dot{\gamma}$is a shear rate, and η is a shear viscosity.

Klass and Martinek [20] introduced a dielectric principle and a water-associated electrical double-layer model, to explain that the critical factor of the ER effect was the molecule in the wet-base ER fluids. The dispersed particles were polarized and distorted when the extra E was applied, and then the water molecules into the dispersion made an attractive bridge between dispersed particles, thereby producing a higher surface tension.

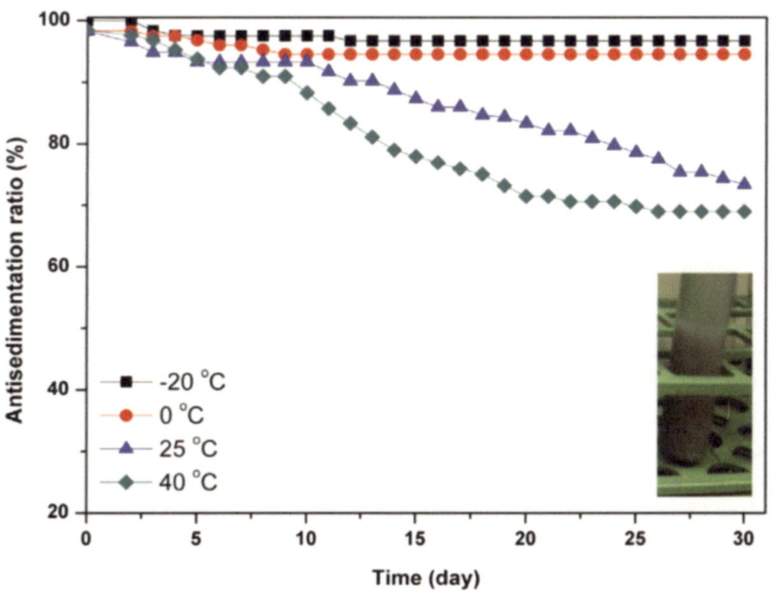

Figure 2.
Antisedimentation stabilities of expanded perlite/silicone oil colloidal dispersions at various temperatures (φ = 10%) [16].

Cho-Choi-Jhon (CCJ) model is an empirical formula with six parameters to explain the ER characteristics in a whole shear rate range [21]. The CCJ equation is given as follows:

$$\tau = \frac{\tau y}{1 + (t_1\dot{\gamma})^{\alpha}} + \eta_{\infty}\left(1 + \frac{1}{(t_2\dot{\gamma})^{\beta}}\right)\dot{\gamma} \quad (3)$$

The first term expresses the decrease in the shear stress at the low shear rate area, and the next term shows a shear-thinning behavior at the high shear rate region. The t_1 and t_2 are time constants, and η_{∞} is a shear viscosity at the infinite shear rate. The exponent α corresponds the shear stress decrease, and b is located between 0 and 1. The CCJ model is observed to be a good indication of the overall behavior of many ER fluids, including the reduced shear stress behavior.

Magnetorheological (MR) fluids consist of dispersed magnetic particles and nonmagnetic fluids such as an hydrocarbon, aqueous carrier fluid, or silicone oil [22, 23]. MR fluids have a Newtonian-like flow behavior with applied magnetic field. However, without an external magnetic field, they show a phase change from a liquid-like to a solid-like due to the formation of a fibrillar structure along the direction parallel to the direction of the applied magnetic field. This behavior is due to the magnetic polarization between the dispersed magnetic particles into the MR dispersion. A hybrid of conducting and magnetic particle-based materials can display an improved dual stimuli–response under electric and magnetic fields [24]. These kinds of fluids also show typical ER and MR behaviors.

2. ER effect and types

On the basis of changes in ER properties of the dispersions, such as viscosity, shear stress, elastic and viscous moduli, and creep recovery, they can be classified as positive or negative ER material depending on their response to the external electric field imposed. In addition to ER effect, depending on the type of applied external stimuli photo, MR and electromagnetorheological effects are also investigated and used in various industrial applications.

2.1 Positive ER effect

An increase in rheological property with applied E is known as the positive ER effect. The fibrillar structure of ER particles is observed in positive ER fluids under E. This dramatic increase in viscosity is due to the change from liquid to solid-like state which is caused by the aggregation of polarized particles into fibrous structures under electric filed. These fibrous structures restrict the motion of the dispersed particles inside the base fluid under the shear flow conditions and result in enhanced viscosity (**Figure 3**).

The positive ER effect is the most common ER phenomenon encountered in the literatures. Conducting polymers (polyaniline, polypyrrole, etc.) [26–28], liquid crystalline polymer/poly(dimethylsiloxane) blends [29], smart polymer/carbon nanotube nanocomposites [30], PPy-tin oxide nanocomposite [31], PAni-coated mesoporous silica [32], core–/shell-structured SiO_2/PPy nanoparticles [33], graphene oxide/PAni composites [34], and biodegradable chitosan/bentonite composites [35] were reported to show positive ER effect.

2.2 Negative ER effect

On the contrary to positive ER effect, for the negative ER effect, the ER properties (i.e., electric field-induced viscosity) are decreased with applied E. the first

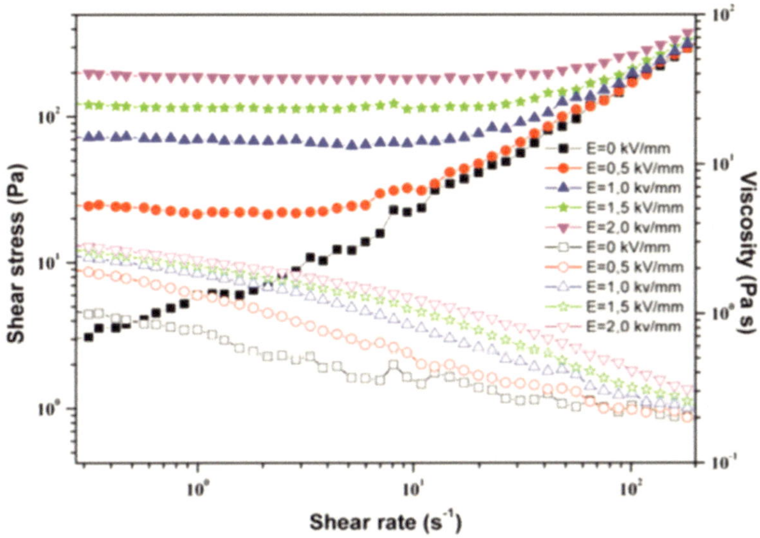

Figure 3.
Change of viscosity and shear stress with shear rate under various E for polyaniline/bentonite composite [25].

negative ER effect was reported by Boissy and his co-workers for the poly(methyl methacrylate) PMMA powder/SO ER fluid [36]. Also, magnesium hydroxide, poly(tetrafluoroethylene) [37], hematite [38], and fumed silica [39] were reported to show the negative ER effect. Urethane-modified polyether liquids exhibit either positive or negative ER effect. Negative ER effect is increased by introducing a branched structure in the polyether main chain or a hard segment at the center of the linear polyether [40]. In a study, colemanite/SO ER fluid showed positive ER effect until E = 1.5 kV/mm and exhibited negative ER effect above E = 1.5 kV/mm. On the other hand, polyindene/colemanite composites dispersed in silicone oil were reported to show negative ER effect with all applied E. This negative ER activity of both the materials was attributed to the electromigration of the particles to one of the electrodes which resulted in phase separation of the dispersion. The presence of Triton X-100 nonionic surfactant converted a negative ER effect to a positive one with improving ER activity. The water/surfactant-bridge formation model was attributed [41]. In another study, the ER properties of urethane-modified polypropylene glycol (UPPG)/PDMS were reported. When the viscosity of UPPG is higher than viscosity of PDMS, ER effect of the suspension is positive. When the viscosity of UPPG is lower than viscosity of PDMS, the ER effect of the suspension is negative [42].

Negative ER effect is explained with two reasons:

- Electrophoresis of particles which migrate to one of the electrodes under the applied electric field leading to phase separation.

- Quincke rotation—Accumulation of charges on the particle surface after the polarization under applied E: the charges on the top and bottom of the surface of particles have the same sign as that of the upper and lower electrodes, and this makes the dispersion unstable and causes the particles to rotate under E.

ER fluids have been used in various industrial areas such as mechanical sensors, damping systems, e-ink, brakes, haptic services, human muscle simulators, and polishing media. Their main advantage is ability to control of mechanical properties with E. These fluids can also exhibit Newtonian fluid characteristics which is the advantage of negative ER fluids [43].

2.3 Photo effect

Light can induce polarization of dispersed particles similar to that of an applied E. Not only positive ER effect but also negative ER effect can be improved by ultraviolet illumination in some ER dispersions. This behavior is called photo-ER effect [44]. Photo-generated carriers are responsible for changing the conductivity of the ER materials and improving their ER response. Photo ER materials show an important relationship between charge mobility and ER activity. Komoda and co-workers studied the photo-ER of TiO_2 nanoparticle dispersions [45]. These dispersions showed positive photo-induced ER effect at low water content, while negative photo-induced ER effect was observed at high water content. Besides, some dye particles (Monastral Green B and copper phthalocyanine), titania nanoparticles, and phenothiazine were reported to show photo ER effect [46, 47].

3. ER and MR materials and their properties

ER materials belong to a family of controllable fluids. They are typically dispersions of solid particles in an insulating liquid. Since the discovery of ER fluids, various types of ER material composition have been proposed [48, 49].

Optical microscopy (OM) studies provide observation of fibrillation of dispersed solid particles under an externally applied *E* which is a characteristic property of ER phenomenon (**Figure 4**).

The dispersed particles have an importance in the ER activity. As the dispersed phase, diverse materials such as polymer blends, gels, biodegradable materials, clays, boron-containing polymers, hybrid nanocomposites, copolymers, ionic liquids, and conducting polymers have been proposed. In recent years, the synthesized ER materials are polystyrene/laponite composites [51], polypyrrole-g-chitosan copolymer [52], polythiophene/polyisoprene composite [53], graphene oxide/titania nanocomposite [54], chitosan/bentonite composite [9], and poly(vinyl chloride)/polyindole composite [55].

The yield stress of MR fluids is generally higher than that of ER fluids. Therefore, MR fluids have a potential for active development. MR fluids show reversible specific properties. They can magnetize under magnetic field and demagnetize in the absence of magnetic field. Magnetite and carbonyl iron (CI) as magnetic particles have been used in the industrial applications of MR fluids [56]. Among the magnetic particles, soft magnetic micron-sized CI particles have an attention for superior MR fluids due to its high saturation magnetization value, suitable size, spherical shape, and low cost [57].

The liquid phase should have high electrical resistivity and hydrophobicity. Liquids such as mineral oil, kerosene, toluene, petroleum fractions, and silicone oil

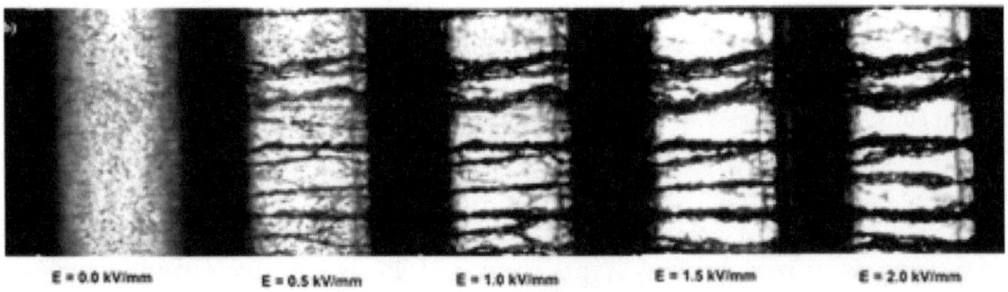

Figure 4.
OM images of fibrillation of dispersed solid nanocube-TiO_2/P_3OT particles [50].

(dimethyl siloxane), dibutyl sebacate, dioctyl phthalate, trioctyl trimellitate, and chlorinated paraffin have been used as dispersing phase [58].

ER fluids are divided into two categories as dry-base ER system and wet-base ER system. The dry-base ER system is anhydrous dispersion and shows ER activity without adding any polar promoter. The wet-base ER system is hydrous dispersion and needs a polar promoter to show ER activity. When these systems compare, dry-base ER system has more attention than wet-base ER system due to its physicochemical characteristics. Many materials such as conducting polymers, biopolymers, and polymer/clay nanocomposites exhibit dry-base ER properties [59, 60].

To enhance the ER effect and to increase the colloidal stability of the dispersions, various types of additives (water, ethyl alcohol, diethylene glycol, surfactants, etc.), called activators, have been added into the dispersions. They most certainly affect the particulate surface and the dispersing liquid.

Surfactants can be added to ER dispersions to improve ER activity and used to tailor dispersion properties. They effect the colloidal stability of dispersions and keep the particles from irreversibly flocculating. Also, they promote the rheological properties of dispersions in the absence an E. While some dispersions show little or no ER effect with adding of water or surfactant, some dispersions display improved ER response with water or surfactant [61, 62]. The adding of water as promoter has several problems such as the low of operation temperature, corrosion, and dispersion conductivity. Some surfactants used in ER dispersions are SDS, CTAB, Triton X, sodium oleate, glycerol, borax, and fatty amines [63–65].

Other additives into the ER dispersions effect the liquid phase by chancing the conductivity. The additives are generally used to increase the rheological characteristics of the dispersions. Some additives used in ER dispersions are water, NaCl, NaOH, acetic acid, amines, oxalic acid, diethylene glycol, glycerine [62]. In a study, the effect of water and surfactant was tested and both in rheological and electrical properties of the dispersions changed by adding small amounts of additives. The yield stress of the ER fluid increased with increasing water content. After the optimal water amount, a reduction in ER activity was observed. This tendency was explained by the water bridge model. On the other hand, the current density of ER fluid increased by adding water and Brij 30. ER activity of Brij 30-activated dispersions decreased under high E conditions. It can be attributed to the formation of surfactant droplets enclosing the particles [66] (**Table 1**).

Antisedimentation stability (or colloidal stability) is one of the important characteristics for the usability of ER dispersions at different environmental conditions. When charged particles in an ER fluid are dispersed, the particle interaction between dispersed particles increases, and rheological properties of ER fluid change. Surface charge of dispersed particles is determined by ζ-potential measurements. Therefore, ζ-potential can be used to determine the colloidal stability of ER fluids.

The density mismatch between the solid and fluid phases in an ER dispersion causes the sedimentation. It can be due to the particle aggregation through van der Walls interaction between the particles and the non-favorable particle-solvent interactions. As a result, the dispersed particles settle down according to Stokes' law [67]. To overcome this traditional problem in ER fluids, several solutions have been suggested.

Surfactants can be used to improve the colloidal stability of the ER fluids. They reduce the sedimentation and aggregation of dispersed particles in an ER fluid via adsorption on the particle surface. This method is still necessary for the long-term stability of ER dispersions. In a reported study, for the synthesis of graft copolymerization with aniline, tailor-made stabilizer was used, and polyaniline particle dispersions with improved stability were obtained [68]. In addition, porosity [69],

Properties	Magnitude
Zero-field viscosity (E = 0 kV/mm)	Low (<1 Pas)
High yield strength	τ = 5 kPa at E = 2 kV/mm
Short response time	<5 ms
Wide working temperature	(−30)–(+125°C)
Low density	d < 1–2 g/cm^3
Chemical resistance	High
Electric field resistance	High
ER materials	Inert, hydrophobic, and cheap
Particle size	<10 μm
Current density	<10 μA/cm^2 at E = 4 kV/mm

Table 1.
Preferred properties of an ideal ER fluid.

particle size [70], and apparent density [52] of dispersed particles play an important role on colloidal stability of the ER fluids.

For MR fluids, sedimentation is an important problem due to the limit of their technological applications. Several approaches to improving the colloidal stability of the dispersed magnetic particles have been suggested. They can be the dispersion stabilizers or additives, viscoplastic carrier medium, and the modification of magnetic particles with either polymers or inorganic coatings [71, 72]. Because of much lower density and relatively good magnetic properties than CI particles, Fe_2O_3 and Fe_3O_4 have been used as MR materials [73].

3.1 Dual response of materials under electric and magnetic fields

Basically the particles used to make MR suspensions cannot be used for ER systems because magnetic particles are more conductive; however, the magnetic particles can be coated with an insulating agent, which renders the suspension applicable in both MR and ER systems. The combined effect of magnetic and electric field produces intensified rheological changes in the suspensions. This synergic effect is termed as electromagnetorheological effect (EMR) [34]. **Figure 5** shows the flow curves of a reported study [74]. As shown, suspension behaves like a Newtonian fluid in the absence of an E or magnetic field, where the shear stress increases linearly with increasing shear rate. On the other hand, it acted as a Bingham fluid under an external E or magnetic field with a yield stress due to the formed particle cluster or chains by polarization forces.

The EMR technique is a new strategy to control the rheological properties of fluid materials. The rheological properties of some fluids can be enhanced with the combination of an electric field and a magnetic field [75]. At this technique, superimposed electric and magnetic fields are applied to a fluid, and the strength/direction of each field can be independently changed during the measurements. Also, the change of magnetic field direction is easier rather than that of the electric field. In a study, a comparison of EMR effects was carried out with the parallel-field and cross-field systems [76]. It was observed that a marked EMR effect was obtained with applied parallel field for the spherical iron dispersion. These EMR effect has been attributed to the different reorientations of the dispersed particles according to field directions. In a reported study, simultaneous impact of electric and magnetic fields on fluids containing a two-component dispersed phase (SiO_2 and CI) was investigated. The fluids showed the strong synergistic effect in the entire investigated range of intensities of electric and magnetic fields. Also, the ferromagnetic particles

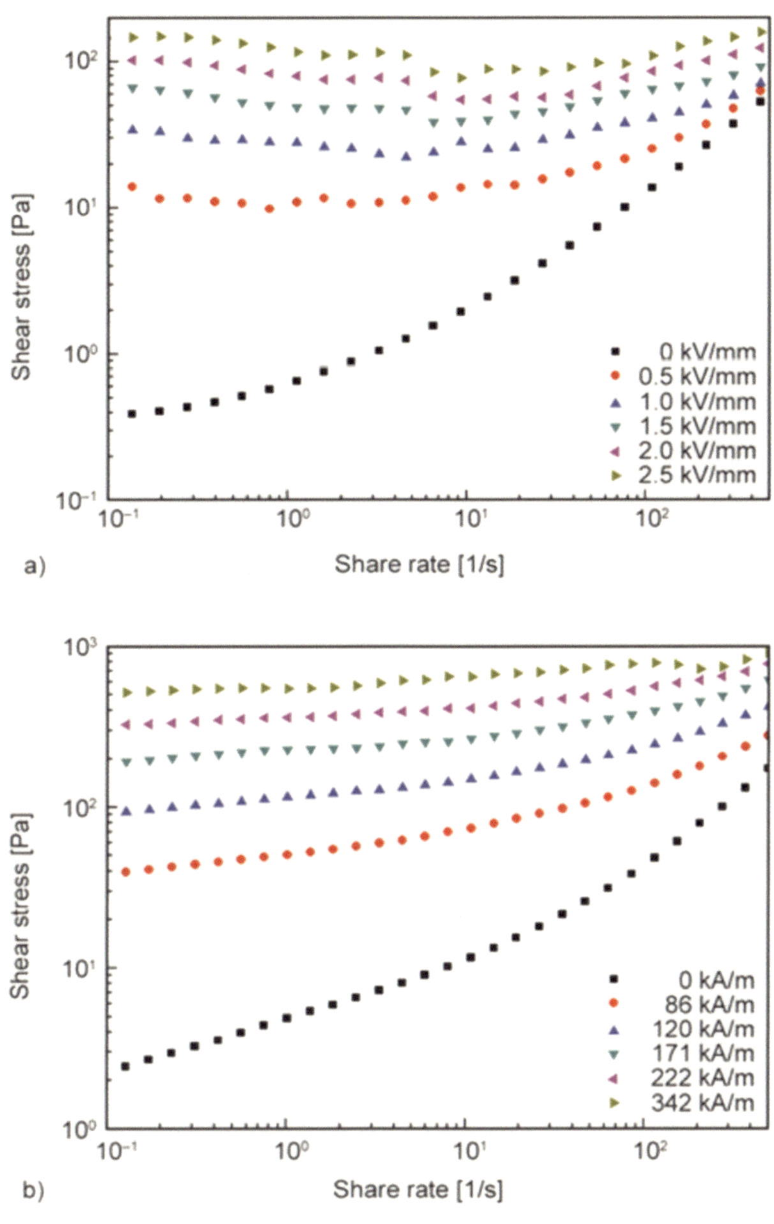

Figure 5.
Flow curves of CI@PANI/SO ER fluid under (a) electric and (b) magnetic fields [75].

(γ-Fe_2O_3) sensitive to the impact of an E was added to the fluids, and an increasing synergistic effect was observed until reaching saturation magnetization [77].

4. Conclusions

Various industrial applications such as the automotive, production sector, and robotic controls are full-potential ER and MR fluid applications. EMR fluids can change the particle interaction and the mechanical properties of dispersed materials. It means that they can allow an increase in the shear stress under two fields. Therefore, EMR fluids can expand the application areas of both ER and MR fluids due to their synergic effects.

Author details

Mehmet Cabuk
Senirkent Vocational School, Isparta University of Applied Sciences, Isparta, Turkey

*Address all correspondence to: mhmtcbk@gmail.com

References

[1] Sedlacik M, Mrlik M, Pavlinek V, Saha P, Quadrat O. Electrorheological properties of suspensions of hollow globular titanium oxide/polypyrrole particles. Colloid and Polymer Science. 2012;**290**:41-48

[2] Hao T. Electrorheological Fluids: The Non-Aqueous Suspensions. Amsterdam: Elsevier; 2005. pp. 83-113

[3] Duff AW. The viscosity of polarised dielectrics. Physics Review. 1896;**4**:23

[4] Winslow WM. Induced fibration of suspensions. Journal of Applied Physics. 1949;**20**:1137

[5] Gawade SS, Jadhav AA. A review on electrorheological (ER) fluids and its applications. International Journal of Engineering Research and Technology (IJERT). 2012;**1**(10):1-8

[6] Ramkumar K, Ganesan N. Vibration and damping of composite sandwich box column with visco elastic/electro rheological fluid core and performance comparison. Materials and Design. 2009;**30**:2981-2994

[7] Oh SY, Kang TJ. Electrorheological response of inorganic-coated multi-wall carbon nanotubes with core-shell nanostructure. Soft Matter. 2014;**10**:3726-3737

[8] Cheng YC, Guo JJ, Xu GJ, Cui P, Liu XH, Liu FH, et al. Electrorheological property and microstructure of acetamide-modified TiO_2 nanoparticles. Colloid and Polymer Science. 2008;**286**:1493

[9] Cabuk M, Yavuz M, Unal HI, Erol O. Synthesis, characterization and investigation of electrorheological properties of biodegradable chitosan/bentonite nanocomposites. Clay Minerals. 2013;**48**:129-141

[10] Zhang WL, Liu YD, Choi HJ. Graphene oxide coated core–shell structured polystyrene microspheres and their electrorheological characteristics under applied electric field. Journal of Materials Chemistry. 2011;**21**:6916-6921

[11] Shen R, Wang X, Lu Y, Wang D, Sun G, Cao Z, et al. Polar-molecule-dominated electrorheological fluids featuring high yield stresses. Advanced Materials. 2009;**21**:4631-4635

[12] Wen W, Huang X, Yang S, Lu K, Sheng P. The giant electrorheological effect in suspensions of nanoparticles. Nature Materials. 2003;**2**:727-730

[13] Shen C, Wen W, Yang S, Sheng P. Wetting-induced electrorheological effect. Journal of Applied Physics. 2006;**99**:106104

[14] Choi HJ, Jhon MS. Electrorheology of polymers and nanocomposites. Soft Materials. 2009;**5**:1562

[15] Wang BX, Zhou M, Rozynek Z, Fossum JO. Electrorheological properties of organically modified nanolayered laponite: Influence of intercalation, adsorption and wettability. Journal of Materials Chemistry. 2009;**19**:1816

[16] Cabuk M. Electrorheological response of mesoporous expanded perlite particles. Microporous and Mesoporous Materials. 2017;**247**:60-65

[17] Slobodian P, Pavlinek V, Lengalova A, Saha P. Polystyrene/Multi-Wall carbon nanotube composites prepared by suspension polymerization and their electrorheological behavior. Current Applied Physics. 2009;**9**:184

[18] Li J, Gong X, Chen S, Wen W, Sheng P. Giant electrorheological fluid comprising nanoparticles: Carbon

nanotube composite. Journal of Applied Physics. 2010;**107**:93507-93512

[19] Engelmann B, Hiptmair R, Hoppe RHW, Mazurkevitch G. Numerical simulation of electrorheological fluids based on an extended Bingham model. Computing and Visualization in Science. 2000;**2**:211-219

[20] Klass DL, Martinek TW. Electroviscous fluids. I. Rheological properties. Journal of Applied Physics. 1967;**38**:67-74

[21] Cho MS, Choi HJ, Jhon MS. Shear stress analysis of a semiconducting polymer based electrorheological fluid system. Polymer. 2005;**46**:11484-11488

[22] Park BJ, Fang FF, Choi HJ. Magnetorheology: Materials and application. Soft Matter. 2010;**6**:5246-5253

[23] Kim MW, Han WJ, Hyun Kim Y, Choi HJ. Effect of a hard magnetic particle additive on rheological characteristics of microspherical carbonyl iron-based magnetorheological fluid. Colloids and Surfaces A: Physicochemical and Engineering Aspects. 2016;**506**:812-820

[24] Noh J, Hong S, Yoon CM, Lee S, Jang J. Dual external field-responsive polyaniline-coated magnetite/silica nanoparticles for smart fluid applications. Chemical Communications. 2017;**53**:6645-6648

[25] Çabuk M, Yesil T, Yavuz M, Hi Ü. Colloidal and electrorheological properties of conducting polyaniline bentonite composite in silicone oil medium. Current Smart Materials. 2017;**2**:1-8

[26] Hiamtup P, Sirivat A, Jamieson AM. Strain-hardening in the oscillatory shear deformation of a dedoped polyaniline electrorheological fluid. Journal of Materials Science. 2010;**45**:1972-1976

[27] Lin DD, Zhang ZJ, Zhao BY, Chen LS, Hu K. Rapid synthesis of porous polyaniline and its application in electrorheological fluid. Smart Materials and Structures. 2006;**15**:1641-1645

[28] Xia X, Yin J, Qiang P, Zhao X. Electrorheological properties of thermo-oxidative polypyrrole nanofibers. Polymer. 2011;**52**:786-792

[29] Kimura H, Koichiro A, Yuichi M, Jun-İchi T, Kiyohito KK. Phase structure change and ER effect in liquid crystalline polymer/dimethylsiloxane blends. Rheologica Acta. 1998;**37**:54-60

[30] Zhang WL, Liu J, Choi HJ. Graphene and graphene oxide composites and their electrorheological applications. Journal of Nanomaterials. 2015;**16**:8

[31] Wei C, Zhu Y, Jin Y, Yang X, Li C. Fabrication and characterization of mesoporous TiO_2/polypyrrole-based nanocomposite for electrorheological fluid. Materials Research Bulletin. 2008;**43**:3263-3269

[32] Noh J, Yoon CM, Jang J. Enhanced electrorheological activity of polyaniline coated mesoporous silica with high aspect ratio. Journal of Colloid and Interface Science. 2016;**470**:237-244

[33] Kim MW, Moon IJ, Choi HJ, Seo Y. Facile fabrication of core/shell structured SiO_2/polypyrrole nanoparticles with surface modification and their electrorheology. RSC Advances. 2016;**6**:56495-56502

[34] Wen W, Sheng P. Two-and three-dimensional ordered structures formed by electro-magnetorheological colloids. Physica B: Condensed Matter. 2003;**338**:343-346

[35] Boissy C, Atten P, Foulc JN. On a negative electrorheological effect. Journal of Electrostatics. 1995;**35**(1):13-20

[36] Young GK, Choi US. Negative electrorheological fluids. Journal of Rheology. 2013;**57**(6):1655-1667

[37] Wu CW. Negative electrorheological effect and electrical properties of a teflon/silicone oil suspension. Journal of Rheology. 1997;**41**:267-275

[38] Erol O, Ramos-Tejada MM, Unal HI, Delgado AV. Effect of surface properties on the electrorheological response of hematite/silicone oil dispersions. Journal of Colloid and Interface Science. **3013**(392):75-82

[39] Minagawa K, Okamura H, Masuda S, Tanaka M, Gohko N, Masubuchi Y, et al. Preparation and property of model homogeneous ER fluids having urethane groups. International Journal of Modern Physics B. 1999;**13**(14n16): 1998-2004

[40] Cetin B, Unal H I and Erol O. The negative and positive electrorheological behavior and vibration damping characteristics of colemanite and polyindene/colemanite conducting composite. Smart Materials and Structures. 2012;**21**(12):125011

[41] Kimura H, Aikawa K, Masubuchi Y, Takimoto J, Koyama K, Uemura T. Positive and negative electro-rheological effect of liquid blends. Journal of Non-Newtonian Fluid Mechanics. 1998;**76**:199-211

[42] Cho MY, Kim JS, Choi HJ, Choi SB, Kim GW. Ultraviolet light-responsive photorheological fluids: As a new class of smart fluids. Smart Materials and Structures. 2017;**26**:054007

[43] Hao T. Electrorheological fluids. Advanced Materials. 2001;**13**(24):1847-1857

[44] Komoda Y, Sakai N, Rao TN, Tryk DA, Fujishima A. Photoelectrorheological phenomena involving Tio_2 particle suspensions. Langmuir. 1998;**7463**(16):1081-1091

[45] Komoda Y, Tata NR, Fujishima A. Photoelectrorheology of TiO_2 Nanoparticle Suspensions. Langmuir. 1997;**13**:1371-1373

[46] Carreira L, Mihajlov V. U.S. Patent 3553708, U.S.A.; 1971

[47] Smith YR, Heermance D, Smith RN. Photo-effects on the viscosity of titania nanoparticle suspensions in conducting and insulating medium. Korea-Australia Rheology Journal. 2016;**28**(1):51-54

[48] Cho MS, Choi HJ, Chin IJ, Ahn WS. Electrorheological characterization of zeolite suspensions. Microporous and Mesoporous Materials. 1999;**32**:233-239

[49] Jang WH, Cho YH, Kim JW, Choi HJ, Sohn JI, Jhon MS. Electrorheological fluids based on chitosan particles. Journal of Materials Science Letters. 2001;**20**:1029-1031

[50] Sever E, Unal HI. Electrorheological, viscoelastic, and creep-recovery behaviors of covalently bonded nanocube-TiO_2/poly(3-octylthiophene) colloidal dispersions. Polymer Composites. 2018;**39**:351-359

[51] Kim YD, Klingenberg DJ. Surfactant-activated electrorheological suspensions. In: Havelka KO, Filisko FE, editors. Progress in Electrorheology. Plenum, New York; 1994. pp. 115-130

[52] Cabuk M, Yavuz M, Unal HI. Colloidal, electrorheological, and viscoelastic properties of polypyrrole-graft-chitosan biodegradable copolymer. Journal of Intelligent Material Systems and Structures. 2015;**26**(14):1799-1810

[53] Puvanatvattana T, Chotpattananont D, Hiamtup P, Niamlang S, Kunanuruksapong R, Sirivat A, et al. Electric field induced stress moduli of polythiophene/polyisoprene

suspensions: Effects of particle conductivity and concentration. Materials Science and Engineering: C. 2008;**28**:119-128

[54] Zhang WL, Choi HJ. Fast and facile fabrication of a graphene oxide/titania nanocomposite and its electro-responsive characteristics. Chemical Communications. 2011;**47**:12286-12288

[55] Koyuncu K, Unal HI, Gumus OY, Erol O, Sari B, Ergin T. Electrokinetic and electrorheological properties of poly(vinyl chloride)/polyindole conducting composites. Polymers for Advanced Technologies. 2011;**23**:1464-1472

[56] Bombard AJF, Knobel M, Alcantara MR. Phosphate coating on the surface of carbonyl iron powder and its effect in magnetorheological suspensions. International Journal of Modern Physics B. 2007;**21**:4858-4867

[57] Ashtiani M, Hashemabadi SH, Ghaffari A. A review on the magnetorheological fluid preparation and stabilization. Journal of Magnetism and Magnetic Materials. 2015;**374**:716-730

[58] Hong CH, Sung JH, Choi HJ. Effects of medium oil on electroresponsive characteristics of chitosan suspensions. Colloid and Polymer Science. 2009;**287**:583-589

[59] Hao T, Xu Z, Xu YJ. Correlation of the dielectric properties of dispersed particles with the electrorheological effect. Journal of Colloid and Interface Science. 1997;**190**:334-340

[60] Kim JW, Choi HJ, Jhon MS. Synthesis and electrorheological characterization of emulsion polymerized SAN-clay nanocomposite suspensions. Macromolecular Symposia. 2000;**155**:229-237

[61] Weiss KD, David CJ. Material aspects of electrorheological systems. Journal of Intelligent Material Systems and Structures. 1993;**4**:13-35

[62] Makela KK. Characterization and performance of electrorheological fluids based on pine oils. VTT Publications 385, Technical Research Centre of Finland, Espoo; 1999

[63] Sudha JD, Sasikala TS. Studies on the formation of self-assembled nano/microstructured polyaniline clay nanocomposite (PANICN) using 3-pentadecyl phenyl phosphoric acid (PDPPA) as a novel intercalating agent cum dopant. Polymer. 2007;**48**:338-347

[64] Unal HI, Sahan B, Erol O. Investigation of electrokinetic and electrorheological properties of polyindole prepared in the presence of a surfactant. Materials Chemistry and Physics. 2012;**134**:382-391

[65] Choi HJ, Jhon MS. Electrorheology of polymers and nanocomposites. Soft Matter. 2009;**5**:1562-1567

[66] Espin MJ, Delgado AV, Plocharsky JZ. Effect of additives and measurement procedure on the electrorheology of hematite/silicone oil suspensions. Rheologica Acta. 2006;**45**:865-876

[67] Uemura T, Minagava K, Takimato J, et al. Opposite electrorheological effects between urethane-based polymers having different terminal groups. Journal of the Chemical Society, Faraday Transactions. 1995;**91**:1051-1052

[68] Armes SP, Aldissi M, Gottesfeld S, Agnew SF. Aqueous colloidal dispersions of polyaniline formed by using poly (vinylpyridine)-based steric stabilizers. Langmuir. 1990;**6**:1745

[69] Yin JB, Zhao XP. Enhanced electrorheological activity of mesoporous Cr-doped TiO_2 from activated pore wall and high surface area. The Journal of Physical Chemistry. B. 2006;**110**(26):12916-12925

[70] Liu F, Xu G, Wu J, Cheng Y, Guo J, Cui P. Electrorheological properties of amorphous titanium oxide particles with different sizes. In: Electro-Rheological Fluids and Magneto-Rheological Suspensions. USA; 2010

[71] Lim ST, Cho MS, Jang IB, Choi HJ, Jhon MS. Magnetorheology of carbonyl-iron suspensions with submicron-sized filler. IEEE Transactions on Magnetics. 2004;**40**:3033-3035

[72] Fang FF, Choi HJ, Seo Y. Sequential coating of magnetic carbonyliron particles with polystyrene and multiwalled carbon nanotubes and its effect on their magnetorheology. ACS Applied Materials and Interfaces. 2010;**2**:54-60

[73] Pu HT, Jiang FJ. Towards high sedimentation stability: Magnetorheological fluids based on CNT/Fe_3O_4 nanocomposites. Nanotechnology. 2005;**16**:1486-1489

[74] Sim B, Chae HS, Choi HJ. Fabrication of polyaniline coated iron oxide hybrid particles and their dual stimuli-response under electric and magnetic fields. Express Polymer Letters. 2015;**9**(8):736-743

[75] Fujita T, Mochizuki J, Lin IJ. Viscosity of electrorheological magneto-dielectric fluid under electric and magnetic fields. Journal of Magnetism and Magnetic Materials. 1993;**122**:29

[76] Koyama K, Minagawa K, Watanabe T, Kumakura Y, Takimoto J-İ. Electro-magneto-rheological effects in parallel-field and crossed-field systems. Journal of Non-Newtonian Fluid Mechanics. 1995;**58**:195-206

[77] Korobko EV, Novikova ZA, Zhurauski MA, Borin D, Odenbach S. Synergistic effect in magnetoelectrorheological fluids with a complex dispersed phase. Journal of Intelligent Material Systems and Structures. 2011;**23**(9):963-967

Section 3

Functional Soft Materials

Chapter 6

An Introduction to Wearable Technology and Smart Textiles and Apparel: Terminology, Statistics, Evolution, and Challenges

Jennifer Xiaopei Wu and Li Li

Abstract

Within the last 6 years, there has been a palpable surge in interest and investment into wearable technology and Smart Textiles & Apparel, manifested both in academia and in the wide array of aspirational products available on the market. Recent market research on wearables also forecasts further growth in the next 3 years. However, as the field becomes increasingly saturated, potential terminology confusion and misconceptions may arise particularly among those in the industry who may not deal directly with the technologies but have interest in applying it to their work. Therefore in an attempt to provide clarity on an increasingly popular field to those perhaps less familiar with it, this chapter delivers an introductory overview of wearable technology and Smart Textiles & Apparel, which clarifies terminology encompassed within the field, reviews recent statistics and maps out how developments have been evolving over time, and assesses some of the challenges confronting the field.

Keywords: wearable technology, smart textiles, apparel, intelligent, electronic textiles

1. Introduction

The field of wearable technology has experienced evident exponential growth in the past 20 years (fueled by the significant increase in mobile device usage and in the expansion of the Internet of Things). Recent market research on wearables also forecasts further growth in the next 3 years [2]. This growth in wearable technology, exemplified in terms of products entering the market and academic research output [1], has been quoted by numerous articles and been used to justify further research potential in the field. However various statistics indicate that the sector of wearable technology on the market represented by Smart Textiles & Apparel, despite the abundance of research output in this sector, continues to be much smaller than that represented by (hard) accessories and devices (e.g., smartwatches and earbuds), which are by far dominating the wearable landscape. Furthermore, various studies show that even with the dominant sector of wearable accessories and devices, there is a high rate of abandonment by users after a short period of time [22].

IntechOpen

Given that "wearable technology" and "Smart Textiles & Apparel" have already gained a great deal of attention and traction in the market and in academic research, there is a tendency for terms to be misused interchangeably or confused; therefore, this review begins by defining and distinguishing terminology used in the scope of Smart Textiles & Apparel. It will then outline the various ways in which Smart Textiles & Apparel have evolved over time by providing examples of innovations in the field. Finally, it will consider some of the practical barriers and challenges in the field.

By mapping out the evolution of Smart Textiles & Apparel as well as its practical barrier and challenges, it will help provide a more realistic sense of the direction the field is headed in the future and subsequently the potential opportunities which need to be addressed in order to propel the field forward—and ultimately toward making a valuable contribution to society.

2. Definitions

As the fields of wearable technology and Smart Textiles & Apparel have expanded and matured over the years, their scope has broadened to encompass increasingly more facets and applications. As a result, a great deal of associated terminology has emerged and some of which have the tendency to be interchangeably or confused with another. Therefore, it is relevant to take a moment to provide clarification and distinction on the quintessential terms of the field (as accepted in academia and the industry, as well as within the bounds of this article).

According to Google Trends (which presents statistics on the amount of interest a search term or topic has garnered online over the years), Web searches on topic of "wearable technology" skyrocketed between 2013 and 2015 by about 500% (see **Figure 1**), which aligns with the proliferation of commercially available wearables, such as the Apple iWatch (which was released in early 2015). However, **Figure 1** also shows that the topic of "wearable technology" had been circulating long (over 20 years) before its surge presence in the last 5 years [3].

The term "wearable technology" (often shortened to just "wearables" and sometimes used interchangeably with "wearable devices" or "smart devices")

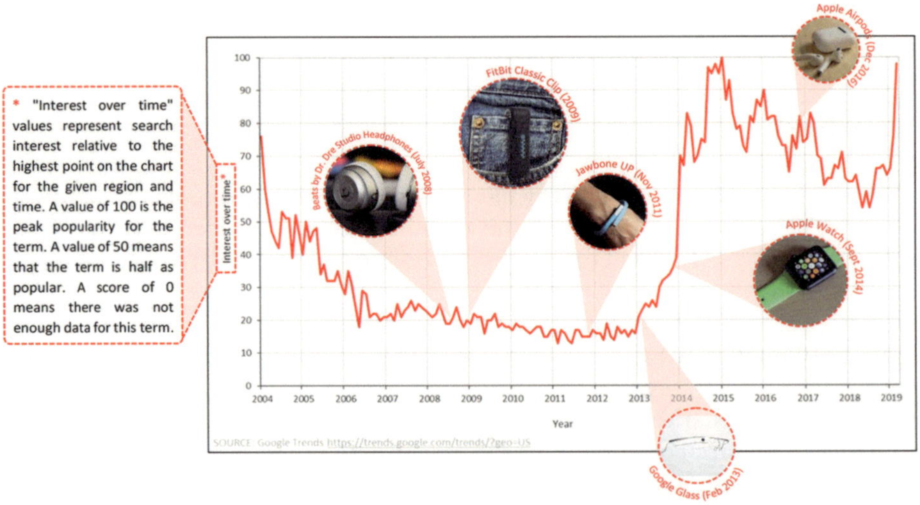

Figure 1.
Graph of Google Trends "Web search interest" on the topic of "wearable technology" over time, with image markers indicating the release dates of the earliest iterations of some "landmark" wearable technology product lines [28–32].

DOI: http://dx.doi.org/10.5772/intechopen.86560

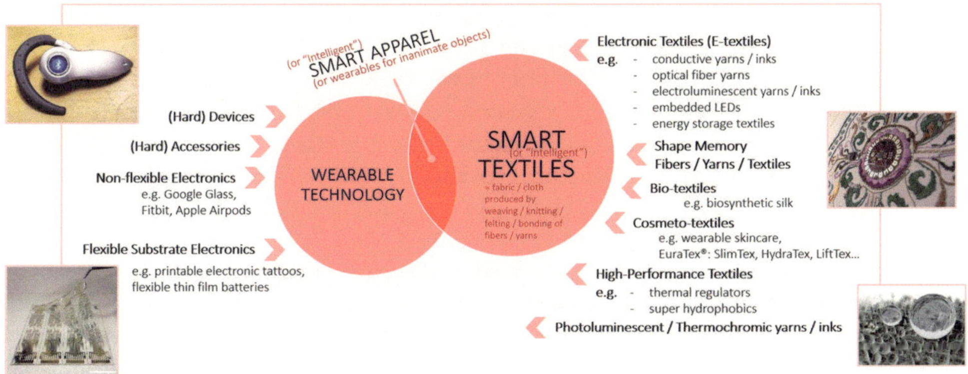

Figure 2.
Diagram to help distinguish and clarify terms within wearable technology and Smart Textiles & Apparel [37, 38, 40, 41].

originated to describe the integration of electronics and computers into clothing or accessories that could be worn comfortably on the body [4], given that the earliest developments, such as the 1999 "Wearable Motherboard," were motivated and enabled by the successive ubiquity of computers, mobile electronics, World Wide Web usage, and big data [5]. However, there are now many examples of advanced "technology" that are also "wearable," but which do not necessarily integrate electronic or computing components—these instead belong to the scope of "Smart Textiles & Apparel." **Figure 2** presents a visual to further help distinguish and clarify wearable technology and Smart Textiles & Apparel as well as provide examples of the types of technologies associated with each area. Therefore, as shown in **Figure 2**, "smart textile" is a category on its own which does not fully overlap with wearable technology, as even though there are wearable technology examples which involve smart textiles (in which case they count as "smart apparel or other wearables inanimate objects"), there are a great deal of smart textiles which are not used toward wearables but toward applications which do not require removal of the textile once applied (such as car interiors or wallpaper). Instead the key distinguishing quality of smart textiles is that they are "soft" materials with flexibility and drapability.

Another pair of terms which are often confused are "smart textiles" and "electronic textiles" ("E-textiles"). Essentially, smart textiles describe a novel category of textiles which have the capability to sense or/and react with or/and adapt to external conditions or stimuli [6], while E-textiles are generally defined as textiles with electronic components or conductive fibers integrated within to give it smart or intelligent functions [7, 8]. Therefore smart textiles are the overarching category which includes E-textiles as one type of smart textile; however, it also includes other types of textiles which exhibit smart or intelligent functions without electronic or conductive elements (see **Figure 2** for examples of the different types of smart textiles).

3. Statistics

To better understand the current and potential future state of Smart Textiles & Apparel, it is helpful to conduct a critical review of the statistics reflecting the progress in the field. One example of such statistics is shown in **Figure 3**, which shows the number of "Smart Textiles & Apparel"—related scholarly publications produced over the years, based on a search via the widely used scholarly publications database

Scopus using the following search criteria: TITLE-ABS-KEY (smart or intelligent or conductive or electronic or sensor) and TITLE-ABS-KEY (textile or apparel or clothing or fashion or knit or yarn or fiber or wearable). **Figure 3** indicates an obvious exponential growth in academic research output in the field of Smart Textiles & Apparel, which appears even more consistent and unwavering than the sporadic surge in Web search topic interest in "wearable technology" from Google Trend statistics (**Figure 1**). However, to gain a more balanced view beyond just academic research, statistics on market presence should also be considered. One source of information which can help gauge and compare the market presence of different forms of wearable technology is the annual attendance audit from the Consumer Electronics Show (CES), a highly prominent conference for innovative technologies being introduced to the market [9]. In particular, data on the number of exhibit attendees representing the "wearables" category was extracted from the annual CES audit summaries of 2012

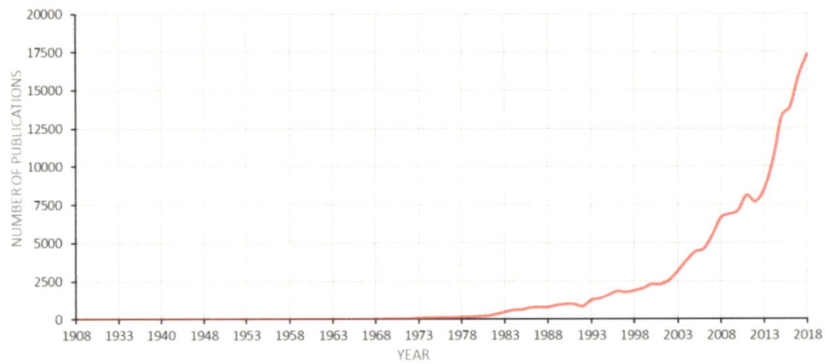

Figure 3.
Number of publications on Smart Textiles & Apparel over time (based on Scopus database search results). Source: Scopus.

8 **Attendee Profiles**

Representation from Top 20 Product Categories

PRODUCT CATEGORY**	ATTENDEES*
Computer Hardware & Software	17,440
Wireless Devices	15,676
Accessories	15,283
Automotive Electronics/Vehicle Technology	14,570
Smart Home/Appliances/Energy Management	13,598
Other Consumer Technology	12,161
Audio	12,106
Wearables	12,059
Telecommunication and Infrastructure	10,049
Video	9,862
Digital Media/Online Media	9,154
Mobile Apps	9,059
E-Commerce	8,965
Audio, High-Performance/High-Resolution	7,692
Augmented Reality/Virtual Reality	7,589
Health and Biotech	7,553
Sensors	6,742
Gaming	6,606
Drones	6,549
3D Printing	5,952

*Attendees = Exhibits Only Attendance; Does not include media or exhibitor personnel
**Attendees may represent multiple product categories

Figure 4.
An example of the CES annual audit summary data on the number of attendees representing the top 20 product categories, from the 2017 CES audit summary.

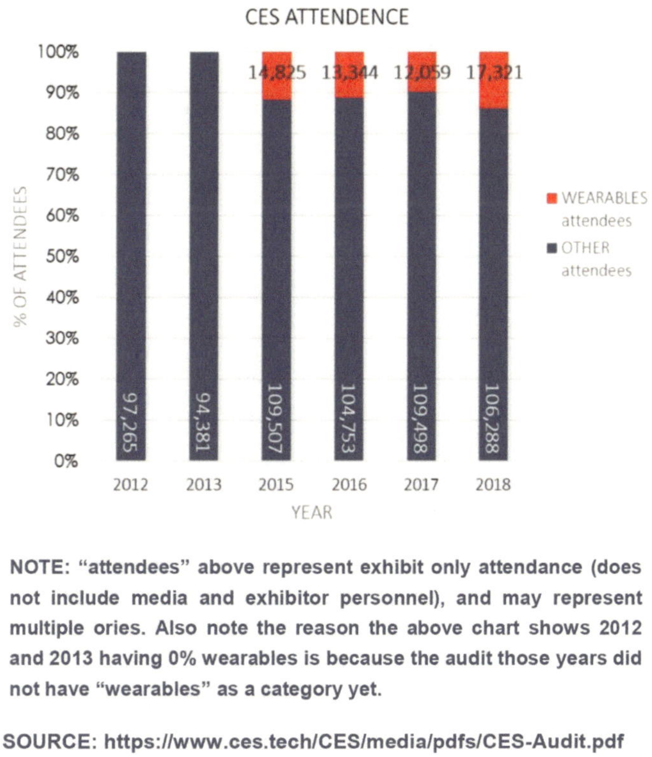

Figure 5.
Numbers of CES exhibit attendees representing the "Wearables" category from 2012 to 2018 (obtained from the annual CES audit summaries).

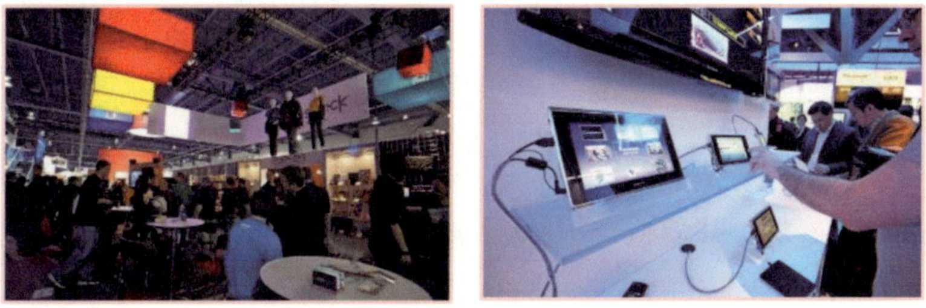

Figure 6.
Photographs taken at the 2011 CES [39].

to 2018 (a sample of this 2017 data is shown in **Figure 4**) and compiled into **Figure 5**. Given that CES is a prominent platform for technologies entering the market, its data compiled in **Figures 5** and **6** validates the growing presence of wearable technology in the market. However, according to statistics from Gartner (**Figure 7**), the share of wearable technology's growing presence occupied specifically by Smart Apparel is much less significant (less than 5% representation), while (hard) accessories (e.g., smart watches, ear-worn devices) represent the dominant category.

Statistics such as that shown in **Figure 7** suggest that Smart Textiles & Apparel still have room to grow in terms of market presence, despite its exponential growth in academic research output. This contrast between its growth in academia and in the market indicates a clear gap and prompts the question of what factors are deterring Smart Textiles & Apparel from being more prominent in the market. This article will start to address this question in the following sections.

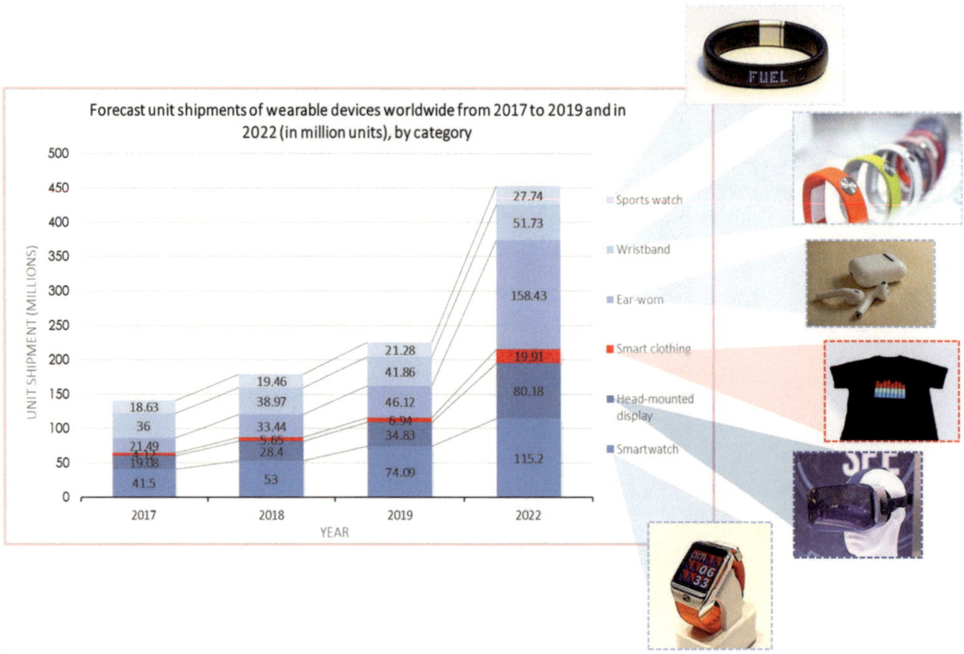

Figure 7.
Global unit shipments of wearable technology by category (2017–2022), alongside image examples. Source: Gartner © Statista 2019 [27, 33–36].

4. Evolution

Understanding how Smart Textiles & Apparel have evolved over the years will help anticipate the direction and opportunities of future growth. Based on an overview of literature, it was deducted that Smart Textiles & Apparel have evolved in three different ways: their degree of integration, their degree of intelligence (or "smartness"), and their degree of self-sufficiency (as summarized in the diagram of **Figure 8**)—each will be explained in the following sections.

4.1 Degree of intelligence (or "smartness")

Smart Textiles & Apparel have evolved in terms of their degree of intelligence (or smartness). This concept is based on the characterization of smart textiles as described by Tao and various other authors in the field, which classifies their degree of intelligence (or "smartness") according to whether they can perform one or more of the following functions: sense, react, and adapt [6, 10, 11].

Traditional textiles (e.g., woven cotton fabrics, knitted wools yarns, etc.), while still functional (i.e., providing warmth, softness, etc.), are latent and do not sense, react, or adapt to external stimuli. The minimum requirement for a smart textile is the ability to sense environmental conditions or an external stimulus, which qualifies it as "passive smart." If it further has the ability to react after sensing, then it qualifies as "active smart." Some examples of this include Grado Zero Espace's shape-memory shirt made from Oricalco fabric which reacts by changing shape based on sensing heat [12] and Aurelie Mosse's intentional use of electro-active light-responsive polymers to create textiles for interiors which sense light and react by changing shape for esthetic effect or for function [13]. Finally, if a smart textile cumulatively has the ability to sense, react, and adapt based on the learned experience from what it sensed and reacted to previously, then it qualifies as "very smart." A conceptual

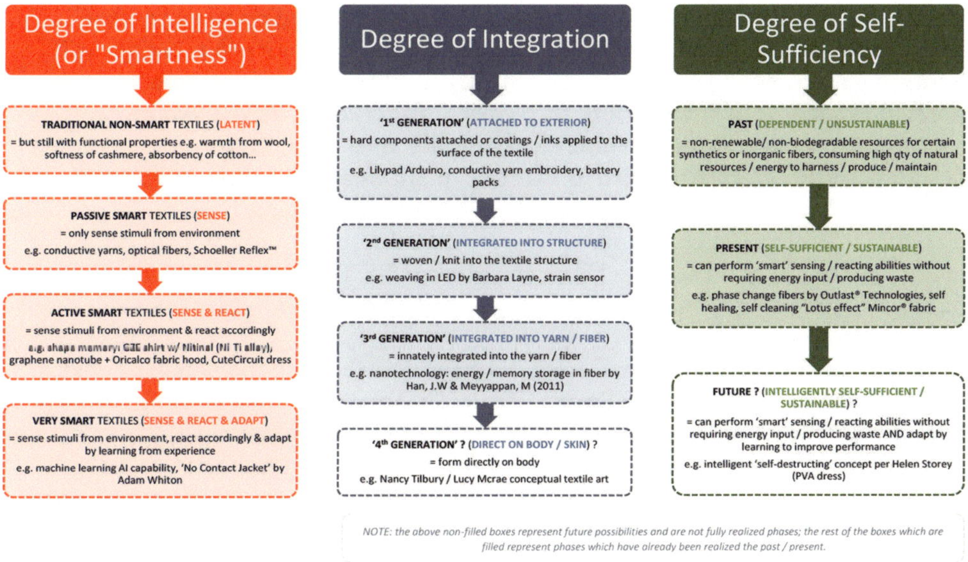

Figure 8.
Flowchart summarizing the evolution of Smart Textiles & Apparel.

example of this is exemplified in the "No-Contact Jacket" created by Adam Whiton and Yolita Nugent, which functions to sense heavy force applied to the jacket (e.g., when wearer is hit by an attacker) and reacts by emitting an electrical chart (i.e., to electrocute the attacker) but also could, in the future, have the potential to integrate machine-learning intelligence to record data on cumulative forces sensed and learn to differentiate between amicable forces (e.g., from a hug or tap) and violent forces, depending on position or amount of force or time of day [13].

See **Figure 8** which presents a visual overview of these definitions of degree of intelligence (or "smartness").

4.2 Degree of integration

Another way Smart Textiles and Apparel have evolved is in terms of their degree of integration. This refers to the extent to which the component/material/substance which performs the "smart" function is embedded into the textile. As Smart Textiles & Apparel have evolved, there has generally been a progression toward a higher degree of embeddedness or integration of the "smart"-functioning component material/substance into the textile. This progression is characterized by Hughes-Riley et al. as three generations: first generation, second generation, and third generation [7]. Although Hughes-Riley describes this progression in the context of E-textiles, it could very well apply to non-electronic smart textiles (e.g., thermochromic inks).

In first-generation smart textiles, the lowest degree of integration, the "smart"-functioning component/material/substance was applied or attached to the surface of the textile; this generation of Smart Textiles Apparel is typically more rigid and bulky and lacks drapability. Some examples of first-generation smart textiles include the use of the "LilyPad Arduino" electronic components on garments [14], printing circuits with conductive ink onto the surface of fabrics, or even embroidering conductive yarns onto fabrics to create conductive pathways. In second-generation smart textiles, the "smart"-functioning component/material/substance is integrated into the textile structure, for example, the weaving or knitting of conductive yarns to form a textile pressure sensor [15], Layne and Orth's work of weaving LEDs, or conductive yarns into the woven textile structure [13]. This generation of Smart Textiles & Apparel is typically more flexible but still tends to be lacking in comfort and versatility. In third-generation smart textiles, the highest degree of integration, the "smart" function is manifested innately as a part of the yarn or fiber and able to exist much more discretely within a textile design without interfering with its esthetic qualities and comfort level. Most examples of this generation of Smart Textiles & Apparel have been made possible by nanotechnology, such as embedding semiconductor systems within fibers for energy harvesting and storage [7] or the Mincor TX TT "lotus effect" coating of nanoparticles on yarns which allow a high level of hydrophobicity to enable a self-cleaning function [16].

The term "fourth generation" used in this article refers to the potential next degree of integration in which Smart Textiles & Apparel are progressing towards but have not yet been fully realized in production and therefore currently exist more as inspiring visionary concepts. One example of a "fourth-generation" concept is the idea of harnessing the innate "smartness" of the nature to create textiles which function "smartly," which is exemplified in the design research of Carole Collette. In her work titled "Suicide Pouf," Collette created cushions which would physically change shape with time as the fibers shed in their natural death process, i.e., relying on the biological process of cell death (apoptosis) to create a dynamic textile object. In "bio-lace," Collette demonstrates an inspiring idea of biologically engineered plants to grow roots which are able to form lace designs on their own [13]. Another visionary concept explored in the work of artists McRae and Tilbury is that of having the human biology or skin to contribute to the creation or modification of what we wear—so not just "smart" functionalities being more integrated into a textile but essentially engineering the human body's biological functions to form its own textiles [13]. Currently, these are rather conceptual ideas, far from mass-market realization. However, they stretch the imagination on the degree of integration that Smart Textiles & Apparel could possibly be achieved in the future.

See **Figure 8** which presents a visual overview of these characterizations of degree of integration.

4.3 Degree self-sufficiency

Smart Textiles & Apparel have also evolved in terms of their degree of self-sufficiency. This refers to the extent to which they can self-sufficiently perform their "smart" functionalities and sustain itself throughout its life cycle, without depending on (non-renewable) energy input or producing (non-biodegradable) waste output. The evolution of degree of self-sufficiency is categorized here as past, present, and future.

The "past" stage refers to Smart Textiles & Apparel which have been made from non-renewable or non-biodegradable raw materials and dependent on non-renewable energy resources to manufacture, to function (e.g., batteries), to maintain (e.g., laundry), and to dispose of. However the growing concern of how polluting the apparel and textile industry and the recent buzz of fashion companies committing to achieving carbon neutrality or employing 100% sustainable materials within the next 10 years [17–19] is motivating a progression toward the "present" stage of the degree of self-sufficiency evolution, which refers to Smart Textiles & Apparel that perform their "smart" (sensing and reacting) functions self-sufficiently (i.e., without requiring a non-renewable energy source input), are self-sustaining (e.g., self-cleaning, self-healing, energy harvesting), and produce limited environmental impact in its manufacture, maintenance, and disposal. The "present" stage factors in sustainability as not only an afterthought but as an intrinsic part of the development and life of Smart Textiles & Apparel. Some examples which fall in this category include the self-cleaning "lotus effect" coating of Mincor TX TT yarn, the use of natural yarns with shape-memory properties which give the textile structure the functionality of becoming more permeable when damp or warm [16], self-healing "green" composites [20], and the light energy absorbing and storing properties of phosphorescent pigments integrated into textiles [13, 21].

One potential "future" direction in the degree of self-sufficiency of Smart Textiles & Apparel could involve an additional element of intelligently adapting (and gradually improving) its self-sufficient smart functions by learning from each experience. So, for example, it could be manifested in the ability of a shape-memory textile to predict when to react without waiting on an external stimulus to trigger it or for a self-healing composite to learn where the weak points in a textile are and produce more self-healing agent in certain areas. These are, once again, only visionary ideas for now; however, with the help of quickly advancing artificial intelligent technologies, its realization could present itself sooner than we think.

See **Figure 8** which presents a visual overview of these characterizations of degree of self-sufficiency.

5. Challenges

Despite burgeoning research into Smart Textiles & Apparel, there remains a great deal of room for improvement in terms of market presence. Furthermore, even with wearable technology (hard) accessories which represent a larger portion of market share, there is the hurdle of low retention and high abandonment rates of such devices less than a year after purchase [22]. Hence below defines some of the challenges contributing to the above disparities and potential opportunities to help tackle those disparities:

- Lack of esthetic appeal: a common feedback on the Smart Apparel options in the market is that they are not fashionable and lack esthetic appeal [23, 24]. What becomes quickly apparent when browsing some of the most exciting wearable innovations at the 2019 Consumer Electronics Show [25] is that they still look more like "techy" electronic devices than something we would recognize as wearable. More and more consumers expect flexibility, comfort, and ease in clothing; however, most smart technologies approved for the market are still relatively bulky and difficult to integrate seamlessly and discretely into traditional clothing materials and styling and dressing habits. (Therefore, as **Figure 7** shows, the smartwatches and ear-worn devices lead in the wearables category at CES, because they have already been adopted as electronic devices that can be worn without causing disruption to or conflict with an outfit). But also, the lack of visibility and accessibility designers in the industry have to Smart Textiles & Apparel technologies is also a barrier to help breed better design esthetics in the field.

- Limited design options: even with wearables that are esthetically well-designed, they tend to come in limited varying options (e.g., in terms of size, colorways, silhouettes, etc.). Currently, the complexity of smart functionalities can limit the design flexibility, so once a successful design is established, it is not easy to vary. This is an issue as today's consumer demand options and flexibility in what they purchase. Therefore, there is a need for Smart Textiles & Apparel technologies to be more versatile, complement different materials, and able to be used in different textile manufacturing processes.

- Lack of assimilation with current supply chain processes/quality standards: the apparel and textile supply chain and quality standards have yet to adapt to the nuances and specific needs of Smart Textiles & Apparel [26]. Because the apparel and textile supply chain is still very much catered to the production of traditional textiles, many suppliers and retailers in the industry might be dissuaded from adopting "smart" technologies due to the time and cost needed to recalibrate processes in the supply chain to manage those "smart" technologies effectively. Furthermore, a great deal of "smart" technologies are not yet developed enough to meet the rigid, long-established quality standards of the textiles and apparel industry or vice versa in that the long-established industry quality standards are not being updated sufficiently to keep up with the incoming "smart" technologies.

- Lack of understanding real human and societal needs: an underlying impression of wearable technology is that it is creating a supply for which there is no real demand (i.e., trying to create a demand). Therefore, in order to make wearables more relevant and desired, there is a need to better understand real human and societal needs so that Wearables can improve in ways to meet those needs. For example, in today's connected world, it is clear that people desire to be connected, but how much so? Where do people draw the line in terms of where they want technology integrated into their everyday lives? When does it become disruptive? These are just some questions that need to be asked to better understand the true needs of humans and society.

- Lack of selling performance and consumer feedback: there is a great deal of statistics available on the quantity of wearable technology and Smart Textiles & Apparel entering or available in the market but limited statistics on how well products sold or on feedback from customers after use. Obtaining such information is important to formulating a clear understanding of real consumer needs. Furthermore, as there are evidently successful wearable technology commercial products out on the market, it would be informative to evaluate what differentiates the successful products from the rest.

6. Conclusion

This review has attempted to provide a brief introductory overview to the now expansive field of wearable technology and Smart Textiles & Apparel, starting by looking at the evident surge in interest in wearable technology in the last 6 years which has been fueled by the release of many "landmark" devices to the commercial market as well as an increasingly mobile-reliant and connected landscape. The exponential growth is also manifested in academic research output in the field, as well as wearable technology products released to the market (which is also forecasted to continue to grow in the future). This chapter also provided clarification on terminology, distinguishing wearable technology and Smart Textiles & Apparel as two separate fields which overlap only in some cases, but also each has their separate unique sub-categories. The chapter then presented the analysis of the progression or evolution of Smart Textiles & Apparel in terms of degree of intelligence, degree of integration, and degree of self-sufficiency. The chapter also takes on a more critical lens to realize the fact that Smart Apparel has a much smaller representation in the wearable technology boom (in comparison with the "hard" wearable accessories and devices such as smartwatches and earbuds which are the dominant categories) as well as the low retention rate of most wearable devices. Given these limitations, the chapter discusses some areas of opportunity for Smart Textiles & Apparel research and product development to enable it to gain more traction and ultimately make a more valuable contribution in the society.

Author details

Jennifer Xiaopei Wu and Li Li*
The Hong Kong Polytechnic University, Hung Hom, Hong Kong

*Address all correspondence to: li.lilly@polyu.edu.hk

References

[1] Van Langenhove L. Smart textiles: Past, present, and future. In: Tao X, editor. Handbook of Smart Textiles. Singapore: Springer Singapore; 2015. pp. 1035-1058

[2] Shirer M, Ramon L, Ubrani J. IDC forecasts steady double-digit growth for wearables as new capabilities and use cases expand the market opportunities. International Data Corporation (IDC); 2019. Available from: https://www.idc.com/getdoc.jsp?containerId=prUS44930019

[3] Google. Google Trends. 2019. Available from: https://trends.google.com/trends/explore?date=all&q=%2Fm%2F06znkfq

[4] Wright R, Keith L. Wearable technology: If the tech fits, wear it. Journal of Electronic Resources in Medical Libraries. 2014;**11**(4):204-216. DOI: 10.1080/15424065.2014.969051

[5] Park S, Jayaraman S. The wearables revolution and big data: The textile lineage. The Journal of The Textile Institute. 2017;**108**(4):605-614. DOI: 10.1080/00405000.2016.1176632

[6] Tao X. Handbook of Smart Textiles. Singapore/New York: Springer; 2015

[7] Hughes-Riley T, Dias T, Cork C. A historical review of the development of electronic textiles. Fibers. 2018;**6**(2):2. DOI: 10.3390/fib6020034

[8] Khan JA. Electronic Textiles. Valley Cottage, NY: Scitus Academics LLC; 2017

[9] Consumer Technology Association (CTA). Consumer Electronics Show (CES). 2019. Available from: https://www.ces.tech/About-CES.aspx

[10] Stoppa M, Chiolerio A. Wearable electronics and smart textiles: A critical review. Sensors. 2014;**14**(7):11957-11992. DOI: 10.3390/s140711957

[11] Tao X. Chapter 1: Smart technology for textiles and clothing—Introduction and overview. In: Tao X, editor. Smart Fibres, Fabrics and Clothing. Sawston, Cambridge: Woodhead Publishing; 2001. pp. 1-6

[12] Pailes-Friedman R. Smart Textiles for Designers: Inventing the Future of Fabrics. London: Laurence King Publishing; 2016

[13] Quinn B. Textile Visionaries: Innovation and Sustainability in Textile Design. London: Laurence King Publishing; 2013

[14] Buechley L, Eisenberg M, Catchen J, Crockett A. The LilyPad Arduino: Using computational textiles to investigate engagement, aesthetics, and diversity in computer science education. 2008. pp. 423-432

[15] Xie J, Jia Y, Miao M. High sensitivity knitted fabric bi-directional pressure sensor based on conductive blended yarn. Smart Materials and Structures. 2019;**28**(3):035017. DOI: 10.1088/1361-665x/aafe89

[16] Kapsali V. Biomimicry for Designers: Applying Nature's Processes and Materials in the Real World. London: Thames & Hudson; 2016

[17] Global Fashion Agenda. 2020 Commitment. 2018. Available from: https://www.globalfashionagenda.com/commitment/

[18] Textile Exchange. Corporate Fiber & Materials Benchmark. 2018. Available from: https://textileexchange.org/pfm-benchmark/

[19] Waste & Resources Action Programme. Sustainable Clothing

Action Plan (SCAP). 2018. Available from: http://www.wrap.org.uk/sustainable-textiles/scap

[20] Kim JR, Netravali AN. Self-healing green composites based on soy protein and microfibrillated cellulose. Composites Science and Technology. 2017;**143**:22-30. DOI: 10.1016/j.compscitech.2017.02.030

[21] Jalal Uddin A. Chapter 9: Novel technical textile yarns. In: Alagirusamy R, Das A, editors. Technical Textile Yarns. Sawston, Cambridge: Woodhead Publishing; 2010. pp. 259-297

[22] Asimakopoulos S, Asimakopoulos G, Spillers F. Motivation and user engagement in fitness tracking: Heuristics for mobile healthcare wearables. Informatics. 2017;**4**(1):5. DOI: 10.3390/informatics4010005

[23] Dunne L. Smart clothing in practice: Key design barriers to commercialization AU—Dunne, Lucy. Fashion Practice. 2010;**2**(1):41-65. DOI: 10.2752/175693810X12640026716393

[24] Housley S. Wearable Tech Has Become Beautiful. So Now What? Technology. 2017. Available from: https://www.wgsn.com/blogs/wearable-tech-has-become-beautiful-so-now-what/

[25] Stoylar B. The Best Wearables of CES 2019. 2019. Available from: https://www.pcmag.com/feature/365914/the-best-wearables-of-ces-2019

[26] Cherenack K, van Pieterson L. Smart textiles: Challenges and opportunities. Journal of Applied Physics. 2012;**112**(9):091301. DOI: 10.1063/1.4742728

[27] Finstad. Pic of a LED T-shirt [Photograph]. 2010. Available from: https://commons.wikimedia.org/wiki/File:LED_T-shirt.jpg

[28] Hook W. Smartwatch [Photograph]. 2015. Available from: https://commons.wikimedia.org/w/index.php?curid=67540374

[29] Keino K. Jawbone UP [Photograph]. 2013. Available from: https://www.flickr.com/photos/djkeino/8667722768

[30] Lisandredu45. Google Glass [Photograph]. 2013. Available from: https://commons.wikimedia.org/wiki/File:Image_GoogleGlass.jpg

[31] LWYang. Airpods [Photograph]. 2017. Available from: https://flickr.com/photos/86688834@N00/34577028814

[32] Parag D. White Beats by Dr. Dre Wireless Headphones [Photograph]. Available from: https://www.pexels.com/photo/beats-depth-of-field-detail-electronics-577769/

[33] Parkes P. Black Steel Nike Fuel Band [Photograph]. 2013. Available from: https://commons.wikimedia.org/wiki/File:Nike_FuelBand.jpg

[34] Pesce M. LeTv LeVR COOL 1 VR Headset [Photograph]. 2016. Available from: https://commons.wikimedia.org/wiki/File:LeTv_LeVR_COOL1_VR_Headset_(24451161536).jpg

[35] Pesce M. Sony Smartband [Photograph]. 2014. Available from: https://commons.wikimedia.org/wiki/File:Sony_Smartband_(12946169423).jpg

[36] Pesce M. Samsung Gear 2 Smartwatch [Photograph]. 2014. Available from: https://commons.wikimedia.org/wiki/File:Samsung_Gear_2_smartwatch_(15059736828).jpg

[37] Stern B. Lilypad Embroidery [Photograph]. 2008. Available from: https://www.flickr.com/photos/bekathwia/2426457410/in/photostream/

[38] Takeda Y, et al. Photograph of Organic CMOS Logic Circuit on a One-Micron-Thick (μm) Substrate. The Total Thickness is Less Than 3 μm. Scale Bar,

25 mm [Photograph]. 2016. Available from: https://commons.wikimedia.org/wiki/File:Organic_CMOS_logic_circuit.jpg

[39] Tech.Co. CES 2011 [Photograph]. 2011. Available from: https://www.flickr.com/photos/techcocktail/sets/72157625757092028/with/5341369273/

[40] Thielicke W. Microscopic Image of a Lotus Leaf with Some Drops of Water, CG of Lotus Effect [Image]. 2007. Available from: https://commons.wikimedia.org/wiki/File:Lotus2mq.jpg

[41] Unknown. A w: Bluetooth Headset for w:mobile Phones [Photograph]. 2004. Available from: https://commons.wikimedia.org/wiki/File:Bluetooth_headset.jpg

Chapter 7

Increased-Value Oxide Powders for Polymeric Fibrous Matrices with Tailored Surfaces for Clothing Wear Comfort: A Review

Narcisa Vrinceanu and Diana Coman

Abstract

This review is dedicated to the area of renewable polymeric matrices, a topic in which the editors aims at conducting, developing, and forming a research-innovation direction for the doctoral studies school. There are two directions envisaging (a) the synthesis of cellulose-derived materials and (b) their application as novel clothing wear materials as an alternative to standard one. The evolution plans aspire to encourage theme-based cooperation between academic environment and industrial media. Thus, the intention is to work on new product development and formulation in a program with strong industry engagement. Specifically, the proposed review will focus closely on applied research and the expansion of the transfer of technology and knowledge in the clothing industries and beyond. Subsequently we propose an interdisciplinary research and development program for material sciences and technology development, meaning, development of new improved ecological comfort performance materials.

Keywords: man-made fibers, oxides, UV reflectance, water barrier, impedance, comfort, UV protection

1. Introduction

This review is dedicated to the area of renewable polymeric matrices, a topic in which the editor develops some research-innovation activities. There are two directions envisaging (a) the synthesis of cellulose-derived materials and (b) their application as novel clothing materials as an alternative to standard one.

The review plans to encourage theme-based cooperation between academic environment and industrial media. Thus, the intention is to work on new product development and formulation in a program with strong industry engagement.

Specifically, the proposed review will focus closely on applied research and the expansion of the transfer of technology and knowledge in the clothing industry and beyond. Subsequently we propose an interdisciplinary research and development program for material sciences and technology development, meaning, development of new improved ecological comfort performance materials.

2. Rationale

Today, it is relevant to develop new research related to the context of environmental and health hazard component.

The exposure of clothing textile materials to wearing plays an important role in the development of adverse health effects.

Washing- and wearing-generated particles are associated with water and air pollution; consequently, pulmonary effects can occur, as shown by both epidemiological and toxicological studies [1–26]. There are important studies, revealing that particles like microfibers, small textiles, and short fibers resulting from washing and wearing of clothing items might contribute to these adverse effects.

Distinct health involvement is connected with the allergenicity of clothing components such as small fibers, small particles derived from wearing of clothing garments, textile wastes, and nondegradable dyes and pigments.

The abovementioned blends have toxic, mutagen features [27–43]. Moreover, respiratory sensitizers, inducing irreversible allergic reactions in the respiratory system, were noticed.

The wastes resulted from washing and wearing of clothing polymeric materials are a key concern point for the European legislation; up to our knowledge, there are no legal demands for the control of banned dyes and pigments having artificial source.

Consequently, some of the approaches relating to non-exhaust particulate matter were highlighted [44]. Researches reveal that the traffic-related emissions are major sources of suspended PM in the urban areas [45–67]. Nevertheless, it can be claimed undoubtedly that data relating to the physical and chemical properties, emission rates, and health consequences of non-exhaust particles derived from specific wearing and washing of textiles are uncertain/far from comprehensive.

Therefore, a logical state of the art touching the validity of the textile waste emission strategies is mandatory, in order to develop practical strategies for reducing the pollutant concentrations.

The review proposes two novel directions tightly bound to the double expertise, *knowledge of the cellulosic polymeric supports treatment against aging process*, gained during the doctoral studies, and a *higher comprehension of synthesis and characterization of nanoporous materials used in clothing wear comfort* derived from textile industry, during my postdoctoral activity. The relevant direction which is the force line stressed by the review is the development of new ecological comfort performance materials.

The development of new comfort performance materials as an alternative sprang from the context of environmental and health hazard within global climate change. During wearing, the clothes could generate wear particles of different chemistries, which are released into the environment, with a different potential of toxicity and mutagenicity.

Moreover, there are various problems associated with synthetics and artificial dyes and pigments, as a consequence of their negative repercussion onto a global and actual background, meaning the life cycle judgment and waste mainframe.

It is well known that manufactured dyes were declared as *taboo*, a lot of research was polarized to test some of their replacements. Thus new formulations appeared for the color of the wear comfort textile materials.

The typical example is carbon fibers proposed as low metallic (or nonmetallic) comfort performant polymeric platforms.

The key point was the idea of synthetic fiber replacement with engineered comfort polymeric materials, having partially the same comfort characteristics, in terms of hygroscopicity, water and air permittivity, thermal, electrical, and sound insulation, and the last but not the least, self-cleaning performance.

There were studies trying to associate synthetic fibers with oxidic nanostructures or to be woven into a single fabric.

Notable scientific research was reported worldwide for the improvement of ecological polymers, for different functions [68].

The review comes up with plant fibers (sisal, jute, hemp, flax) utilized as pillared structures for various polymers, replacing synthetic fibers (glass, Kevlar, carbon, etc.). Due to their biodegradability and sustainability, cellulosic fibers are tremendous backup which can be utilized as basics in a particular variety of polymeric composite applications (e.g., kenaf and betel nut fibers) [44, 68–86].

Scientific reports made important state of the art, by summarizing data regarding an extensive review concerning the employment of natural-sourced fibers as fundamentals in polymer-based composites, with a certain target on their self-clean/photocatalytic behavior [87].

The relevant requirements, which these new engineered comfort composites with self-cleaning purposes should fulfill, are the following:

- Acceptable values of the comfort performance coefficient
- Stability at higher temperatures

The most extremely serious aspect is the one regarding the thermal stability of the plant fibers. In terms of thermal stability, in comparison with the most valued aramid fibers, decaying between 400 and 450°C, these thermal conditions are diminished.

This fact happens because of the temperature at the contact surface can locally exceeds several hundred degrees, during intense breaking.

The question is whether the natural plant fibers can be used for application at such higher temperatures.

The research plan highlights the concept of using three main approaches to increase plant fibers in thermal stability:

- The augmentation of elemental composition of cellulose, by eliminating the secondary components, like hemicellulose and lignin, by an alkaline method
- The enhancement of thermal cohesion of the novel comfort performant polymeric composites by employment of montmorillonite (MMT) emulsion, having attributes of inorganic natural clay, containing SiO_4 tetrahedral sheets arranged into a two-dimensional network structure, thus granting thermal protection [88]
- The direct growth of pure and doped nanostructured ZnO coating onto the abovementioned fibrous polymeric matrix as active photo cleaned/stain repellent material in high engineered comfort products

In other words, the research review proposes the creation of some polymeric platforms, whose place remains an important question.

Based on the work experience in the metal oxide and other photoactive materials for coating area, a novel approach to the direct growth of pure and doped nanostructured ZnO coating onto different kinds of polymeric supports as active photocatalyst material in comfort performance is proposed.

By means of governing the nucleation dynamics, an alternative to modify the expansion process of oxidic nanostructures occurs. Thus, the enlargement of oxidic nanostructures is possible in a competitive mode at low temperature, by employment of physical approaches such as laser ablation, plasma vapor deposition, nonaqueous solution growth, and solgel and spray deposition.

The abovementioned methodology grants the control of distinct polymeric supports for oxidic nanostructure nucleation, having the convenience of a considerable larger surface than usual, low interfaces conducting to an augmented surface in volume ratio of the active self-cleaned polymeric supports.

3. Impact

The expected impact of the *first research direction* is to make available new scientific knowledge regarding comfort performant polymeric supports able to consistently reduce stain repellence of them, since there is a niche in this field.

The expected outcomes will be a benefit to the know-how in the domain of original textile platforms with comfort behavior, since green-chemistry engineered substrates are in their inception. This review can make accessible an extensible territory of technological employments, considering aspects like expenditure and low environmental impact of the comfort polymeric platforms, with potential commercialization in the future. Since the research plan targets air purification in fact, the achieving of the fundamental objectives will have a strong environmental, social, and economic impact.

The replacement in the model formulation of mineral fillers and fibers (e.g., glass fibers); metals, such as copper, lead, and tin; antimony trisulfide; and aramid pulp, presently used in comfort performance materials, by renewable natural fiber-reinforced polymeric matrix composites (NF: cotton, flax, or sisal fibers), will lead to:

- Thanks to their biodegradability, the production of comfort textile components with potentially complete recyclability will be sustainable (zero waste at the end of the life cycle).

- An achievable lower weight will mean a decrease of gas emission in vehicles enhancing the quality of life, due to the low density of the NF.

- More safety during fiber managing and a longer life for processing tools will be provided by the lower abrasiveness and friendly handling of natural fibers as reinforcing elements.

- Positive health effects.

- Eco-friendly, lack of toxicity.

- Low cost and weight, with potential perspective commercialization.

- Better ratio properties/weight, compared to glass fibers on the expense of lower structural properties.

- Mechanical properties identical to those of traditional polymeric composites, reduced friction at wear and tear, high geometric stability of the manufactured parts, and good insulation characteristics.

The innovative engineering of advanced eco-pad was obtained from green-chemical, nonmetal materials, entirely fulfilling the most relevant promoting tendency of the modern textile comfort architecture.

Having into consideration, the other *research force line*, the desired impingement was to create reachable new scientific overview regarding self-cleaning polymeric platforms, in order to capture/block/entrap the chemical compounds resulted from the mineralization/degradation of organic matters, where this kind of expertise is unpredictable, vague.

4. Objectives

4.1 For the novel comfort performant polymeric platforms

The novelty/originality of the proposed chapter consists in the application of life cycle assessment (LCA) and eco-design methods to evaluate and optimize comfort performance material formulations, through a comprehensive consideration of resource consumption and bio-toxicity. In the above context, and given the variety and complexity of non-exhaust emission sources, the statement proposes some specific objectives, obviously derived from the fundamental objective of the research plan.

- To obtain new ecological enhanced comfort performant polymeric platforms (eco-pads). The role of plant fibers in these comfort performance composites will be studied in relation to formulation/engineering, comfort performance materials, and comfort performance material surfaces.

- To optimize the product/process: processability and performances of the composites. This aspect requires an optimization of the matrix components and the processing procedures. It is compulsory to optimize composition (percentage of polymers and fiber; type and amount of additives and filler), procedure (mixing, extrusion; treatment of fibers), and process conditions (pH, temperature).

- To apply the extension evaluation method of wearing comfort performance materials, which is an effective tool for the ranking/selection of comfort performance materials, based on some defined criteria as follows: performance, physical properties, costs of raw materials, wearing and washing effectiveness, and thermal stability.

- To create a small-scale prototype pad, integrating the best comfort performance material behavior to reduce the emission of particles derived from wearing and washing, up to 50%.

- To create a database for environmental impacts and bio-toxicity of comfort performance materials, to evaluate and optimize the comfort performance material formulations through a comprehensive consideration of resources consumption and bio-toxicity.

4.2 For the self-cleaning polymer-based platforms

The originality of this plan consists in the use as growth support of polymeric materials to increase photocatalytic/self-cleaning active material surface, the use of the simple chemical approach for direct nucleation of ZnO nanostructures, as well as the use of ZnO-based polymeric supports as self-cleaning/photocatalyst in an innovative prototype platform.

1. In order to design nanostructured oxide-based coatings on polymeric substrates with restrained sizes, favorable homogeneity, and durability using as standard nucleation methods, diverse chemical simple ways (aqueous chemical growth, nonaqueous solution growth, and sol gel), utilizing cheap precursors, and feasibility for potential economic purposes.

2. To detect the excellent polymeric platform configuration, morphology, and attributes in each polymeric layer and to obtain an outside surface to volume ratio for operative membrane, in the case of particular chemical nucleation path.

3. To accomplish materials with high hydrophobicity/stain repellency during wearing.

4. To create a small-scale prototype reactor, integrating the best self-cleaning active nanooxide-based polymeric supports, in terms of sustainability, feasibility, and cost-effectiveness for future scale-up.

5. Methodology

5.1 For the novel comfort performant polymeric platforms

1. **Preparation of eco-friendly comfort performance composites with tailored comfort performance formulations**, given the expertise on cellulosic supports behavior:

- Cellulose-based fibers in composites formulation

- *Cotton composite*: polyester resin reinforced with montmorillonite-coated cotton fibers

- *Banana/pineapple composite*: epoxy resin reinforced with banana/pineapple fibers

  Sample preparation:

  - Extract fibers.

  - Prepare epoxy and hardener.

  - Prepare mold.

  - Fabricate composite.

- *Jute/ramie* fibers in mixture with powdered nut shells as natural and biodegradable fillers in non-asbestos organic (NAO) comfort performance material composites (graphite will be replaced with nut shell)

- *Jute composites*: Jute/polyester-clay (montmorillonite) composite, jute vinyl ester composite, jute epoxy composite, and jute/polypropylene/polystyrene thermoplastic composite; montmorillonite-coated jute fiber; and reinforced polyvinyl chloride (PVC) film composites

- *Sisal composites*: Montmorillonite-coated sisal/polyester composite, sisal/epoxy composite, sisal urea/formaldehyde composite, and sisal polystyrene/polypropylene composite

- *Flax composite*: Montmorillonite-coated flax polyester composite, flax/epoxy composite, flax/polystyrene composite, and flax/polypropylene composite

- *Hybrid composite* with different volume fraction of glass and bamboo fibers reinforced with an epoxy polymer

- *Wool-polyester* resin composites

- *Aramid pulp and natural fibers*

A special attention will be given to chemical-free polymeric supports using agricultural wastes as a source of raw materials (coconut shell). In terms of economical wastes, utilization is not a favorable solution, but the result might be environmentally evaluated/quantified.

Method of production: dry wastes, grind, make different-sized sieves, and use a compression molding machine.

Methods of characterization: compressive strength test, flame resistance, water and oil (SEA 20/50) absorption, and wear rate.

Methods for the making of natural fiber composites [89]:

a. Filament winding

b. Hand lay-up/spray consisting of two processes

Characterization methods of eco-friendly comfort performance composites by analytical techniques will be combined to study the topography, microstructure, chemical composition, and thermal stability of the comfort performance polymeric platforms, including SEM for morphology, EDX for elemental analysis, XRD, TGA, EMA, XRF, and profilometry.

- *Identifying the compounds in the raw components of wearing clothing items by* atomic absorption spectrometry (AAS) and atomic emission spectrometry (AES) with inductively coupled plasma (ICP) spectrometry

- *Physical properties*: bulk density and water absorption properties

- *Tensile properties*: test direction, fracture stress, fracture strain, and Young's modulus

- *Mechanical properties*: tensile strength and elongation break

- *Thermal properties by DTA and TGA analyses*

Measurements of comfort performance polymeric platforms by:

- *Comfort performance materials* tester according to the SAE J661 recommendation [88–97]
- Constant speed comfort performance material test machine
- Comfort performance material friction assessment and screening test (FAST) machine
- Single-ended full-scale dynamometer
- *Abrasive wear test* (*Taguchi method*)

Characterization of microstructures of comfort performant material surfaces/layers belonging to the wear comfort performant composites:

a. Comfort performance material surface morphology by scanning electron microscopy with energy dispersive X-ray microanalysis and electron microprobe analysis, and light microscopy (LM)

Phase analysis:

b. X-ray powder diffraction methods

c. X-ray fluorescence spectrometry

d. Thermal gravimetric analysis

Identification of:

1. Dynamic response: the structural morphology of comfort performant polymeric layers modifies according to the distinct superficial points and across specimen's thickness.
2. Comfort performant polymeric platforms status correlation: the engineering of comfort performant polymeric layers is in strong interrelation with parameters, such as interval of temperature, duration, and thermal behavior such as degradation and readjustment 11.
3. Elemental composition dependency: the content of polymeric surface and bulk varies; however, bulk amount determines the comfort performance of polymeric material thickness.

Evaluation of:

- Wearing performance characteristics
- Thermal stability and comfort wear properties
- Wearing system performance assessment
- Relationships between comfort performance materials and compositions of the comfort performance materials

5.2 For the self-cleaning polymeric platforms

1. Direct nucleation of pure and doped ZnO/TiO_2 nanostructures within various polymeric matrices/membranes, by chemical approaches (aqueous chemical growth, nonaqueous solution growth, and sol gel)

    - The nucleation method will be enhanced, in terms of type of substrate. The selection of the materials was made in terms of potential support for the photocatalytic ZnO layer (natural fiber polymeric support, synthetic fibrous matrices, and mineral fiber fibrous matrices) with different textures. Another criterion was the envisaged application; thus, materials with low optical absorption in UV-Vis region of the light spectrum were preferred.

    - ZnO growth parameters optimization for each support.

    - Mn, Ag, Cu, Ni, V, and S, with various concentrations, will be used as dopants in ZnO lattice in order to enhance its photocatalytic activity in the visible region of electromagnetic spectrum and tailor morphology.

2. Systematic study of structural, chemical, and physical properties of pure and doped ZnO nanostructured material grown on various polymeric supports and growth optimization with respect to the substrate and dopant

    - Pure and doped ZnO material characterization: study of topology of the coating involving optical and electron microscopy SEM, TEM, AFM, structure, composition and stoichiometry XRD, EDX, FT-IR, Raman, BET, and optical properties using UV-Vis spectroscopy

    - Coated polymeric support characterization testing: coating quality inspection using microscopic techniques, optical properties in the case of transparent textile substrates, adherence studies, washing tests, and hydrophilic behavior evaluation

3. Evaluation of the photocatalytic activity for each material on each substrate followed by optimization of the growth technique with respect to the photocatalytic efficiency

    - Preliminary photocatalytic tests employing wearing rubbishes from the inside of the polymeric layers.

    - Photocatalytic activity against various particles generated by washing and wearing. The tests employ various light sources. The tests were done for one contaminant at a time in synthetic air and water. The pollutant depletion was monitored for a reasonable time, and the results will be compared to establish which pair of photocatalyst substrate has the highest activity.

4. Integration of the optimum coated polymeric material in a novel laboratory small-scale prototype of a photocatalytic reactor, which may be the subject of a patent application

    - Design of small-scale laboratory urban air decontamination prototype reactor to host the best-engineered photocatalytic active material

- Execution of the designed prototype reactor for air decontamination and integration of the best photocatalytic active material
- Testing the prototype reactor for air decontamination and optimization with respect to application and aesthetical factors

Acknowledgements

Project financed from "Lucian Blaga" University of Sibiu Research Grants LBUS-IRG-2016-02.

Author details

Narcisa Vrinceanu* and Diana Coman
"Lucian Blaga" University of Sibiu, Faculty of Engineering, Department of Industrial Machines and Equipment, Romania

*Address all correspondence to: vrinceanu.narcisai@ulbsibiu.ro

References

[1] Abramov OV, Gedanken A, Koltypin Y, Perkas N, Perelshtein I, Joyce E, et al. Pilot scale sonochemical coating of nanoparticles onto textiles to produce biocidal fabrics. Surface and Coatings Technology. 2009;**204**:718-722

[2] Aizenberg J, Albeck S, Weiner S, Addadi L. Journal of the American Chemical Society. 1993;**115**:11691-11697

[3] Allen NS, Edge M, Corrales T, Childs A, Liauw CM, Catalina F, et al. Ageing and stabilization of filled polymers: An overview. Polymer Degradation and Stability. 1998;**61**:183-199

[4] Anderson RB. Modifications of the brunauer, emmett and teller equation. Journal of the American Chemical Society. 1946;**68**(4):686-691

[5] Anitha S, Brabu B, Thiruvadigal DJ, Gopalakrishnan C, Natarajan TS. Optical, bactericidal and water repellent properties of electrospun nano-composite membranes of cellulose acetate and ZnO. Carbohydrate Polymers. 2012;**87**:1065-1072

[6] Applerot G, Perkas N, Amirian G, Girshevitz O, Gedanken A. Coating of glass with ZnO via ultrasonic irradiation and a study of its antibacterial properties. Applied Surface Science. 2009;**256**:S3-S8

[7] Asif MH, Elinder F, Willander M. Electrochemical biosensors based on ZnO nanostructures to measure intracellular metal ions and glucose. Journal of Analytical and Bioanalytical Techniques. 2011;**S7**:1-9

[8] Ates ES, Unalan HE. Zinc nanowire enhanced multifunctional coatings for cotton fabrics. Thin Solid Films. 2012;**520**:4658-4661

[9] Becheri A, Dürr M, Nostro PL, Baglioni P. Synthesis and characterization of zinc oxide nanoparticles: Application to textile as UV-absorbers. Journal of Nanoparticle Research. 2008;**10**:679-689

[10] Bender ML, Komiyama M. Cyclodextrin Chemistry. Berlin: Springer; 1978

[11] Valentin TI, Vrinceanu N, Pachiu C, Bucur S, Pascariu P, Rusen L, et al. Nanostructured ZnO based materials for biomedical and environmental applications. In: Dinca V, Suchea M, editors. Functional Nanostructured Interfaces for Environmental and Biomedical Applications. 1st ed. Elsevier; 2019. p. 436. ISBN: 9780128144015. https://www.elsevier.com/books/functional-nanostructured-interfaces-for-environmental-and-biomedical-applications/dinca/978-0-12-814401-5

[12] Biscan J. Electrokinetic data: Approaches, interpretations and applications. Croatica Chemica Acta. 2007;**80**(3-4):357

[13] Bledzki AK, Gassan J. Composite reinforced with cellulose based fibers. Progress in Polymer Science. 1999;**24**:221

[14] De Boer JH. The Dynamical Character of Adsorption. 2nd ed. Oxford: Clarendon Press; 1968. pp. 200-219

[15] Borcia C, Borcia G, Dumitrascu N. Surface treatment of polymers by plasma and UV radiation. Romanian Journal of Physics. 2011;**56**:224-232

[16] Boulter PG. A review of emission factors for road vehicle non-exhaust particulate matter. In: TRL Report PPR065 TRL Limited: Wokingham; 2005

[17] Brayner R, Ferrari-Iliou R, Brivois N, et al. Toxicological impact

studies based on *Escherichia coli* bacteria in ultrafine ZnO nanoparticles colloidal medium. Nano Letters. 2006;**6**:866

[18] Broasca G, Campagne C, Farima D, Ciocoiu M, Iorgoaea M, Vrinceanu N. Advanced materials and technologies. In: The International Scientific Conference UGALMAT 2011; Galati, Romania; 2011. p. 175

[19] Broasca G, Suchea MP, Ciocoiu M, Farima D, Vrinceanu N, Campagne C, et al. Novel approach regarding zeta potential variation as a consequence of a woven fabrics antibacterial finishing treatment. "Tekstil ve Konfeksiyon" (Journal of Textile and Apparel). 2013;**23**(1):43-48, ISSN: 1300-3356

[20] Broasca G, Borcia G, Dumitrascu G. Narcisa Vrinceanu: "Characterization of ZnO coated polyester fabrics for UV protection". Applied Surface Science. 2013;**279**:272-278, ISSN: 0169-4332

[21] Brunauer S, Deming LS, Deming WE, Teller E. On a theory of the van der Waals adsorption of gases. Journal of the American Chemical Society. 1940;**62**(7):1723-1732

[22] Buksek H, Luxbacher T, Petrinic I. Zeta potential determination of polymeric materials using two differently designed measuring cells of an electrokinetic analyzer. Acta Chimica Slovenica. 2010;**57**:700-706

[23] Cao G. Nanostructures & Nanomaterials: Synthesis, Properties and Applications. Imperial College Press; 2008

[24] Chen B, Peng X, Wang J, Sun S. Multi-scale hybrid numerical simulation of the growth of high-aspect-ratio nanostructures. Computational Materials Science. 2008;**4**:201

[25] Chen CG, Khobaib M, Curliss D. Epoxy layered silicate nanocomposites. Progress in Organic Coating. 2003;**47**:376-383

[26] Chibowski E, Mittal KL. Contact Angle, Wettability and Adhesion. Vol. 2. Utrecht: VSP; 2002. p. 265

[27] Chow JC, Watson JG, Egami RT, Franzier CA, Lu Z, Goodrich A, et al. Evaluation of regenerative-air vacuum street sweeping on geological contributions to PM10. Journal of the Air and Waste Management Association. 1990;**40**(8):1134-1142

[28] Christopher PW, Bruce HJ. Early silica cementation. Sedimentary Research. 1988;**58**

[29] Corrales T, Catalina F, Peinado C, Allen NS, Fontan E. Photooxidative and thermal degradation of polyethylenes interrelationship by chemiluminescence, thermal gravimetric analysis and FT-IR data. Journal of Photochemistry and Photobiology A. 2002;**147**:213-224

[30] Didane N, Giraud S, Devaux E. Thermal and fire resistance of fibrous materials made by PET containing flame retardant agents. Polymer Degradation and Stability. 2012;**97**:1083-1089

[31] Drexler EK. Engines of Creation: The Coming Era of Nanotechnology. New York, US: Anchor Books; 1986

[32] Dura'n N, Marcato Purity D, De Souza GIH, Alves OL, Esposito E. Antibacterial effect of silver nanoparticles produced by fungal process on textile fabrics and their effluent treatment. Journal of Biomedical Nanotechnology. 2007;**3**:203-208

[33] Eita M, Wagberg L, Muhammed M. Spin-assisted multilayers of poly(methyl methacrylate) and zinc oxide quantum dots for ultraviolet-blocking applications. ACS Applied Materials & Interfaces. 2012;**4**:2920-2925

[34] El-Rafie HM, El-Rafie MH, Zahran MK. Green synthesis of silver anoparticles using extracted from marine macro algae. Carbohydrate Polymers. 2013;**96**:403-410

[35] Faal N, Farzaneh F. Study of heating effect on specific surface area, and changing optical properties of ZnO nanocrystals. Applied Surface Science. 2006;**17**(3):231-234

[36] Fan Q, John J, Ugbolue SC, Wilson AR, Dar YS, Yang Y. Nanoclay-modified polypropylene dyeable with acid and disperse dyes. AATCC Review. 2003;**3**(6):25

[37] Feynman RP, Leighton RB, Sands M. The Feynman Lectures on Physics. Vol. 1. Reading, MA: Addison-Wesley; 1964. pp. 2-12

[38] Fujihara S, Sasaki C, Kimura T. Crystallization behavior and origin of c-axis orientation in sol-gel-derived ZnO. Applied Surface Science. 2001;**180**:34

[39] Furuta T, Kusuya Y, Neoh TL, Rehmann L, Beak SH, Yoshii H. Kinetic analysis and evaluation of controlled release of D-limonene encapsulated in spray-dried cyclodextrin powder under linearly ramped humidity. Journal of Inclusion Phenomena and Macrocyclic Chemistry. 2006;**56**:107

[40] Gardner TJ, Messing GJ. Magnesium salt decomposition and morphological development during evaporative decomposition of solutions. Thermochimica Acta. 1984;**78**:17

[41] Garvey CJ, Parker IH, Simon GP. On the interpretation of X-ray diffraction powder patterns in terms of the nanostructure of cellulose fibres. Macromolecular Chemistry and Physics. 2005;**206**:1568-1575

[42] Gawas UB, Mojumdar SC, Verenkar VMS. Synthesis and characterization of Co0.8Zn0.2Fe_2O_4 nanoparticles Journal of Thermal Analysis and Calorimetry. 2009;**96**:49-52

[43] Gawas UB, Verenkar VMS, Mojumdar SC. Synthesis and characterization of Co0.8Zn0.2Fe_2O_4 nanoparticles. Journal of Thermal Analysis and Calorimetry. 2011;**104**:879-883

[44] Gerlofs-Nijland ME, Dormans JAMA, Bloemen HJT, Leseman DLAC, Boere AJF, Kelly FJ, et al. Toxicity of coarse and fine particulate matter from sites with contrasting traffic profiles. Inhalation Toxicology. 2007;**19**:1055-1069

[45] Gowri VS, Saxena M. Protection of bamboo surfaces by CNSL based coatings. Journal of Chemical Technology. 1997;**14**:145-149

[46] Gonsalves LR, Mojumdar SC, Verenkar VMS. Synthesis and characterization of Co0.8Zn0.2Fe_2O_4 nanoparticles. Journal of Thermal Analysis and Calorimetry. 2011;**104**:869-873

[47] Grigoriu AM. Cercetări în domeniul compusilor de incluziune ai ciclodextrinelor si al derivatilor acestora cu aplicatii în industria textilă [Ph. D. diss.]. Iasi; 2009

[48] Grigoriu AM, Luca C. Noi aplicații ale ciclodextrinelor în industria textilă. Iasi: Performantica Publishing House; 2010

[49] Grigoriu AM, Luca C, Vrinceanu N, Ciolacu F. Inclusion compounds of monochlorotriazinyl-β-cyclodextrin grafted on a paper support. Cellulose Chemistry and Technology. 2011;**45**(3-4):177-183

[50] Gu F, Wang SF, Lu MK, Zhou GJ, Xu D, Yuan DR. Structure evaluation and highly enhanced luminescence of Dy3+-doped ZnO nanocrystals by Li+ doping via combustion method. Langmuir. 2004;**20**:3528

[51] Guggenheim EA. Application of Statistical Mechanics. Oxford: Clarendon Press; 1966. pp. 186-206

[52] Gui XC, Wei JQ, Wang KL, et al. Carbon nanotube sponges. Advanced Materials. 2010;**22**:617

[53] Guo L, Yang S, Yang C, Yu P, Wang J, Ge W, et al. Synthesis and Characterization of Poly(vinylpyrrolidone)-Modified Zinc Oxide Nanoparticles. Chemistry of Materials. 2000;**12**:2268-2274

[54] Guo L, Campagne C, Perwuelz A, Leroux F. Zeta potential and surface physico-chemical properties of atmospheric air-plasma-treated polyester fabrics. Textile Research Journal. 2009;**79**(15):1371-1377

[55] Hatch KL, Osterwalder U. Garments as solar ultraviolet radiation screening materials. Dermatologic Clinics. 2006;**24**:85-100

[56] He Y, Yang B, Cheng G. Oxidative coupling of methane over Li/MgO: Catalyst and nanocatalyst performance. Catalysis Today. 2004;**98**:595

[57] He JH, Kunitake T, Nakao A. Facile in situ synthesis of noble metal nanoparticles in porous cellulose fibers. Chemistry of Materials. 2003;**15**:4401-4406

[58] He JH, Kunitake T, Nakao A. Facile fabrication of composites of platinum nanoparticles and amorphous carbon films by catalyzed carbonization of cellulose fibers. Chemical Communications. 2004;**4**:410-411

[59] Hingorani S, Pillai V, Kumar P, Multani MS, Shah DO. Microemulsion mediated synthesis of zinc-oxide nanoparticles for varistor studies. Materials Research Bulletin. 1993;**28**:1303-1310

[60] Huang MH, Wu YY, Feick HN, Tran N, Weber E, Yang PD. Catalytic Growth of Zinc Oxide Nanowires by Vapor Transport. Advanced Materials. 2001;**13**:113

[61] Huang Z, Zheng X, Yan D, et al. Toxicological effect of ZnO nanoparticles based on bacteria. Langmuir. 2008;**24**:4140

[62] Hughes LJ, Britt GE. Compatibility studies on polyacrylate and polymethacrylate systems. Journal of Applied Polymer Science. 1961;**5**:337

[63] Hunter DRJ. Zeta Potential in Colloid Science: Principles and Applications. London: Academic Press; 1981

[64] Ibanez JA, Forte J, Hernandez A, Tejerina F. Streaming potential and phenomenological coefficients in Nuclepore membranes. Journal of Membrane Science. 1988;**36**:45

[65] Israelachvili J. Intermolecular and Surface Forces. second ed. London: Academic Press Ltd.; 1992

[66] Jacobasch HJ, Baubock G, Schurz J. Problems and results of zeta-potential measurements on fibers. Colloid and Polymer Science. 1985;**263**:3

[67] Jacobasch HJ, Schurz J. Characterization of polymer surfaces by means of electrokinetic measurements. Progress in Colloid and Polymer Science. 1988;**77**:40

[68] Behera JK. Synthesis and characterization of nano-particles [M. Tech thesis]. NIT: Rourkela

[69] Jeevani J. Nanotextiles-A broader perspective. Nanomedicine and Nanotechnology. 2011; Available from: http://omicsonline.org/nanotextiles-a-broader-perspective-2157-7439.1000124.pdf

[70] Jiadao W, Ying Y, Darong C. Research progress on the ultra hydrophobic surface topography effect. Chinese Science Bulletin. 2006;**51**:2297

[71] Jiang SQ, Newton E, Yuen CWM, Kan CW. Chemical silver plating and its application to textile fabric design. Journal of Applied Polymer Science. 2005;**96**:919-926

[72] Jiang L, Yao X, Li HX, Fu YY, Chen L, Meng Q, et al. All-carbon

electronic devices fabricated by directly grown single-walled single-walled carbon nanotubes on reduced graphene oxide electrodes. Advanced Materials. 2010;**22**:376

[73] Jiang ZH, Zhong Y, So CS, Hse CY. Rapid prediction of wood crystallinity in *Pinus elliottii* plantation wood. Journal of Wood Science. 2007;**53**:449

[74] Jiugao Y, Jingwen Y, Baoxiang L, Xiaofei M. Preparation and characterization of glycerol plasticized-pea starch/ZnO-carboxymethylcellulose sodium nanocomposites. Bioresource Technology. 2009;**100**:2832-2841

[75] Jongnavakit P, Amornpitoksuk P, Suwanboon S, Ratana T. Surface and photocatalytic properties of ZnO thin film prepared by sol-gel method. Thin Solid Films. 2012;**520**:5561-5567

[76] Joshi M, Bhattacharyya A. Nanotechnology: A new route to high-performance functional textiles. Textile Progress. 2011;**43**:155-233

[77] Kathirvelu S, D'Souza L, Dhurai B. UV protection of textiles using ZnO nanoparticles. Indian Journal of Fibre and Textile Research. 2009;**34**:267-273

[78] Kekkonen V, Hakola A, Kajava T, Sahramo E, Malm J, Karppinen M, et al. Self-erasing and rewritable wettability patterns on ZnO thin films. Applied Physics Letters. 2010;**97**:044102

[79] Kelly FJ, Fussell JC. Air pollution and airway disease. Clinical and Experimental Allergy. 2011;**41**(8):1-13. DOI: 10.1111/j.1365-2222.2011.03776.x

[80] Khedr MGA, Abdel Haleem SM, Baraka A. Degradation of poly(ether sulfone)/polyvinylpyrrolidone membranes by sodium hypochlorite: insight from advanced electrokinetic characterizations. Journal of Electroanalytical Chemistry. 1985:184

[81] Kittelson DB. Engines and nanoparticles: A review. Journal of Aerosol Science. 1998;**29**:575-588

[82] Kuhns H, Etyemezian V, Landwehr D, MacDougall C, Pitchford M, Green M. (2001). Testing reentrained aerosol kinetic emission from roads (TRAKER): A new approach to infer silt loadings on roadways. Atmospheric Environment. 2001;**35**:2815-2825

[83] Kuhns H, Etyemezian V, Green M, Hendrickson K, McGown M, Bartond K, et al. Vehicle-based road dust emission measurement—Part II: Effect of precipitation, wintertime road sanding, and street sweepers on inferred PM10 emission potentials from paved and unpaved roads. Atmospheric Environment. 2009;(37):4573-4582

[84] Kwon YJ, Kim KH, Lim CS, Shim KB. Characterization of ZnO nanopowders synthesized by the polymerized complex method via an organochemical route. Journal of Ceramic Processing Research. 2002;**3**:146-149

[85] Lee HJ, Yeo SY, Jeong SH. Antibacterial effect of nanosized silver colloidal solution on textile fabrics. Journal of Materials Science. 2003;**38**:2199-2204

[86] Leroy P, Tournassat C, Bizi M. The influence of surface conductivity on the apparent zeta potential of TiO_2 nanoparticles. Journal of Colloid and Interface Science. 2011;**356**(2):442-453

[87] Li J, Yang YX, Zha F, Lei ZQ. Facile fabrication of superhydrophobic ZnO surfaces from high to low water adhesion. Materials Letters. 2012;**75**:71-73

[88] Li Y, Duan G, Cao B, et al. Superhydrophobicity of 2D ZnO ordered pore arrays formed by solution-dipping template method. Journal of Colloid and Interface Science. 2005;**287**:634

[89] Li YQ, Fua SY, Mai YW. Preparation and characterization of transparent ZnO/epoxy nanocomposites with high-UV shielding efficiency. Polymer. 2006;**47**:2127-2132

[90] Liu JP, Qu SC, Zeng XB, Xu Y, Gou XF, Wang ZJ, et al. Fabrication of ZnO and its enhancement of charge injection and transport in hybrid organic/inorganic light emitting devices. Applied Surface Science. 2007;**253**:7506-7509

[91] Liufu S, Xiao H, Li. Investigation of PEG adsorption on the surface of zinc oxide nanoparticles. Powder Technology. 2004;**145**:20-24

[92] Lorne B, Perrier F, Avouac JP. Streaming potential measurements. Journal of Geophysical Research. 1999;**104**(17):17857-17877

[93] Lu HF, Fei B, Xin JH, Wang RH, Li L. Fabrication of UV-blocking nanohybrid coating via miniemulsion polymerization. Journal of Colloid and Interface Science. 2006;**300**:111-116

[94] Lu L, Farris TN, Chandrasekar S. Sliding microindentation wear particles: Spheres in grinding swarf. In: Dowson D, Taylor CM, Childs THC, Godet M, Dalmaz G, editors. From the Cradle to the Grave. Leeds, UK Elsevier; 1992. pp. 257-263

[95] Luca C, Grigoriu AM. Antimicrobial polymer films functionalized with cyclodextrins. Revista de Chimie. 2006;**57**:248

[96] Ma XF, Chang PR, Yang JW, Yu JG. Preparation and properties of glycerol plasticized-pea starch/zinc oxide-starch bionanocomposites. Carbohydrate Polymers. 2009;**75**:472-478

[97] Mani G, Fan Q, Ugbolue SC, Eiff IM. Effect of nanoparticle size and its distribution on the dyeability of polypropylene. AATCC Review. 2003;**3**(1):22